METAVERSE

www.royalcollins.com

METAVERSE

KEVIN CHEN

Books Beyond Boundaries

ROYAL COLLINS

METAVERSE

KEVIN CHEN
Translated by Daniel McRyan

First published in 2023 by Royal Collins Publishing Group Inc.
Groupe Publication Royal Collins Inc.
550-555 boul. René-Lévesque O Montréal (Québec) H2Z1B1 Canada

ISBN: 978-1-4878-1023-8

To find out more about our publications, please visit www.royalcollins.com.

Foreword

As computing power continues to grow and technical innovations such as high-speed wireless communication networks, cloud computing, blockchain, virtual engines, VR/AR, and digital twins gradually converge, generation Z and millennials will live a digital life unimaginable in the past – as more things are digitized, human-computer interaction becomes infinitely close to or even better than human-to-human interaction, and massive digital assets are created, mined, traded and consumed. The digital economy keeps rising and is now dominant, marking a drastic change in consumer attention.

A hybrid world prototype, where virtuality and reality are intertwined and every-thing is connected, was born with the help of Web 3.0 technology. In the ideal world of Web 3.0, people no longer bother to distinguish between physical reality and digital virtuality. More importantly, they will want all their friends, personal belongings, and experiences to be virtually connected. This hybrid world – where virtuality and reality are intertwined while everything is connected – is the metaverse.

In fact, as early as 1992, the concept of the metaverse appeared in the novel *Snow Crash* by Neal Stephenson. It mainly refers to the integration of the three modes of physical reality, augmented reality, and virtual reality, in a shared cyberspace. The 2018 movie *Ready Player One* showcased this idea more vividly: it created an Oasis scene where players could freely explore, entertain, and live in the virtual world through VR equipment. From *Snow Crash* to *Ready Player One*, this concept has gradually become more resolute and understandable.

It was on March 10, 2021 that the metaverse actually became part of reality. The sandbox game platform Roblox was the first company to include this metaverse concept into its prospectus. Its IPO on the New York Stock Exchange turned out to be a great success on the first day – its market value exceeded 40 billion US dollars, stunning both technology and financial circles. Thereafter, the metaverse went viral, attracting enormous attention from science and technology, business, and governments.

It is worth mentioning that the contemporary metaverse in 2021 has gone beyond that conveyed in *Snow Crash* in 1992. The metaverse now, having absorbed the achievements of multiple revolutions of digital technology, shows mankind the possibility of building a holographic digital world that is parallel and integrated with the conventional physical world. It also promotes breakthroughs in traditional philosophy, sociology, and the humanities.

This book aims to succinctly introduce the origins of metaverse from a technical perspective and to predict the development thereof. The metaverse connects numerous digital technologies. The reconstruction of computing power builds the metaverse; 5G consolidates the foundation of the communication network; artificial intelligence becomes its "brain," meaning to control it in the future; the digital twin is its tentacle from the future; blockchain creates the economic system; virtual technology represented by VR/AR/MR is the critical path … This is how the metaverse connects the dots into lines and lines into planes. And this, in turn, is how it turns science fiction into reality.

We may not be able to accurately describe how it will look, but judging from the front-end signs and development tendencies that have been presented so far, the future metaverse will be a new era of the comprehensive interconnection of things, accurate expression of objects, precise perception of human beings, and intelligent interpretation of information. What we are going to feel will be an intricate network system of super-large scale, infinite expansion, rich layers, and harmonious operations. This is generated by the information exchange network, resulting in a brand-new civilization landscape integrating the real world and the digital world.

The virtual world changes from simulating and replicating the physical world to extending and expanding it, then reacts to it, and eventually blurs the boundaries between the two. This is an important vision for humanity's future. Certainly, like the growth process of any living organism, before the metaverse becomes a mainstream fact, there is a mountain of arduous work to complete in order to pave its way. When technologies mature, and the content ecology is complete, the grand idea of the metaverse will come true. When that happens, society will have a new meaning.

KEVIN CHEN
October 8, 2021 in Hong Kong, China

Contents

CONTENTS

CONTENTS

METAVERSE

CHAPTER 1

Dawn of the Metaverse

From the novel *Snow Crash* 29 years ago to the 2018 film *Ready Player One*, the metaverse remains just a fictional world in sci-fi works. People fantasize about having a completely virtual shared space where they can freely travel between the virtual and the real world, have virtual identities, and build friendship circles and economic systems synchronously.

Despite much discussion in recent years on parallel worlds, virtual humans, cloud games, digital twins, and extended reality (XR), only now has the metaverse become the real talk of the town. In a hallmark event for the industry, Roblox entered the capital market in March 2021. From sci-fi to reality, the door to the metaverse has been truly opened, and there is a window to look forward to. The occurrence of all this gives credit to the elevated position and promotion of multiple technologies. This has clearly outlined the development path for the metaverse.

1.1 Beginning

The metaverse first appeared in *Snow Crash*, the third book by science fiction author Neal Stephenson, published in 1992. Conceptually, the word is composed of meta and verse. Meta means beyond, while verse stands for the universe. Together, they refer to the idea of being "beyond the universe," that is, an artificial space that runs parallel to the real world.

The story of *Snow Crash* takes place in Los Angeles in the early 21st century. In the future world imagined in the novel, the government distributes most of its power to private entrepreneurs and organizations, and national security is entrusted to mercenary soldiers. Businesses compete with each other to attract more resources, while the government's remaining power is only used for some tedious work. The prosperity and stability of society have nothing to do with it. Most of the government's land is divided among private individuals, who in turn establish personal territory. Under this circumstance, people discover a drug called *Snow Crash*, which is, in fact, a computer virus that can spread not only on the Internet, but also in real life, causing system crashes and brain malfunction.

In this way, Stephenson did not create the Internet familiar to us, but a three-dimensional digital space closely connected with society – the metaverse. In this space, people in the real world who are geographically isolated can communicate and entertain themselves through their respective avatars. The adventure of the protagonist, Hiro, unfolds in this metaverse based on information technology.

Chapter 5 of *Snow Crash* paints a clear image of the metaverse where Hiro lives:

"He is not seeing real people, of course. This is all a part of the moving illustration drawn by his computer according to specifications coming down the fiber-optic cable. The people are pieces of software called avatars. They are the audiovisual bodies that people use to communicate with each other in the Metaverse. Hiro's avatar is now on the Street, too, and if the couples coming off the monorail look

over in his direction, they can see him, just as he's seeing them. They could strike up a conversation: Hiro in the U-Stor-It in L. A. and the four teenagers probably on a couch in a suburb of Chicago, each with their own laptop. But they probably won't talk to each other, any more than they would in Reality. These are nice kids, and they don't want to talk to a solitary crossbreed with a slick custom avatar who's packing a couple of swords."

In the novel, Hiro's job is to deliver pizza to the Mafia that has taken control of the American territory. When he is not working, he enters the metaverse. In virtual reality, people act as avatars of their own design, engaging in mundane activities such as chatting and flirting, as well as extraordinary tasks such as sword fighting and mercenary espionage.

Like the Internet, the metaverse is a collective and interactive endeavor. It is always on yet is not under any individual's control. Just like a video game, people inhabit and control characters that move through space. The main stem and world order of the metaverse follow the Global Multimedia Protocol formulated by the Association for Computing Machinery. Developers need to obtain a development license for land before they can build small alleys, skyscrapers, parks and various things that violate the physical laws in reality.

After *Snow Crash*, well-known film and television productions such as *The Matrix* in 1999, *Sword Art Online* in 2012, and *Ready Player One* in 2018 have dramatized people's interpretation and imagination of the metaverse.

Compared with *Snow Crash*, *The Matrix* incorporates extensive philosophical elements, such as existentialism, structuralism, fatalism, and nihilism. Similarly, the film also builds a metaverse different from the real world. In the metaverse of *The Matrix*, human beings are connected to a virtual neural network – the matrix – through a jack. The virtual neural network can simulate reality and let the brain hallucinate through electromagnetic signals. Consequently, it takes full control over real human beings, from body to soul.

Thereafter, human genes exist in the computer programmers while thoughts exist in the matrix. The matrix observes various human activities to achieve self-learning, resulting in upgrades and transformations to artificial intelligence machines. The matrix has undergone six upgrades; at last, it comes to the most critical upgrade, which is to enable the robots to love in order to observe the significance of human love to the evolution of life. This is a dangerous plan that could lead to the total self-destruction of the Matrix-built AI world. The protagonist Neo is the one chosen for this upgrade.

Having swallowed the red pill from Morpheus, Neo wakes up from the virtual world and finds himself naked inside an embryo incubator in an enormous machine factory. There are tubes on his body that connect him to the virtual world, where he has lived for over 20 years without knowing the truth. Neo looks around and finds himself in a vast life base. He is just one of the thousands of ignorant souls. The terrifying manipulator is constantly watching, and new life is doomed to have no soul the moment it is born. A game of truth and falsehood begins.

In the 2018 film *Ready Player One* directed by Spielberg, the metaverse is effectively known as the Oasis. The story takes place in the year 2045, in which the real world is on the brink of chaos, and collapse is imminent. A virtual reality universe created by the genius James Halliday attracts great public attention, and people hope to find redemption in the Oasis. There, they can race cars and go on adventures. And the somatosensory clothing and VR equipment enable players to feel the sensory stimuli in the virtual world.

There is a prosperous city and an independent economic system in the virtual universe. People can be anyone and do anything there. A loser in reality can be a superhero in the Oasis. The Oasis is a mirror world parallel to the real world, where people escape their disappointing reality, leave behind their frustration, and find themselves again.

From *Snow Crash* to *The Matrix*, and to *Ready Player One*, the concept of metaverse has gradually become clear. It is a virtual world born out of the real world, parallel to it, interacting with it, and always online.

1.2   From Meta to Metaverse

Profound Meaning of Meta

Meta comes from the Greek prefix μετά, which means after, beyond, above, or between. And these meanings are noticeable in the words "metaphysics" and "meta-economy," though they are not commonly used today. Aristotle wrote about metaphysics in the 4th century BC, discussing the nature of reality that people may explore after studying the physical world. In *The Metamorphosis*, a novel by Franz Kafka, meta refers to the transformation from an old to a new form.

元 (yuan, meaning meta) in Chinese is an ancient character that was first seen during the Shang Dynasty in bone and bronze inscriptions. It resembles a man standing side-on with a protruding head. Therefore, its literal meaning is head; the head is the highest part of the human body and holds a paramount function. It is also used to represent the origin of all things in the world, meaning the root. Slowly, 首 has replaced yuan to refer to the head, and the extended meaning of yuan is used more used. *Shuo Wen Jie Zi* (Explanation and Study of Principles of Composition of Characters) explains that yuan means the beginning of everything, and *Luxuriant Dew of the Spring and Autumn Annals* by Dong Zhongshu concisely and expressively defines it as "the root of all things."

Moving forward, the modern concept of meta began with the metamathematics proposed by David Hilbert in 1920. Generally, metamathematics refers to the use of mathematical techniques to study mathematics itself: it is scientific thinking or knowledge that uses math as human consciousness and cultural objects. This self-referential meaning became the core of most versions of meta-anything.

With the advent of the LISP (list processing), meta began to have technical connotations. LISP, born in 1958, is the second oldest and still widely used high-end programming language. (Only the programming language FORTRAN predates it by a year.) LISP is the most popular programming language in artificial intelligence (AI) research, partially because it can do meta programming – writing or manipulating other programs (or itself) as their data, or finishing work that should have been done earlier. As LISP became popular, many keyboards designed for LISP programmers even included a meta key.

In 1968, ten years after LISP, John Lilly applied the concept of metaprogramming to humans in *Programming and Metaprogramming in the Human Biocomputer: Theory and Experiments*, and Timothy Lilly, an LSD expert in the 1960s, once called it "one of the three most important ideas of the 20ᵗʰ century." Lilly believed that our environment is constantly programming us. In experiments with a potent semi-hallucinogen and LSD, Lilly believed people should be allowed to modify their own programs.

In 1979, Douglas Hofstadter published *Gödel, Escher, Bach: An Eternal Golden Braid in collaboration with Basic Books*. The book adopts the earlier usage of meta in metamathematics and metaprogramming as a prefix for self-reference.

The meaning of meta in popular culture has therefore been tentatively defined: when we talk about something as "meta," we're doing self-referentially. The meaning of self-referentially talking about something meta makes it a stand-alone adjective used to describe a type of self-reflection and self-observation.

The meaning of self-observation in meta is widely seen in art. For the simplest example, the protagonist of a book writing a book or that of a movie making a movie is called meta. Some works have adopted the form of meta exaggeratingly. For example, the film *Birdman* tells the story of an actor who plays a superhero in the same-name film, and tries to restart his acting career on the stage.

The usage of meta in popular culture can be described with a formula: meta + B = B about B. When we add the prefix meta to a word, for example, metacognition is cognition about cognition.

Universe About the Universe

Although the meaning of the meta is clarified, the market holds different views on the concept of the metaverse. The following list is a compilation of some expert opinions on the matter:

Tim Sweeney, CEO of Epic Games, believes it is a real-time 3D medium for large-scale participation never seen before. It enables us to enjoy real-time social interaction in a virtual world, and it has a fair economic system. All creators can participate, earn money, and be rewarded. While Fortnite, Minecraft, and Roblox exhibit some of the metaverse's features, they're nowhere near it.

David Baszucki, CEO of Roblox believes that the metaverse includes eight characteristics: identity, friendship, immersion, low latency, diversity, omnipotence, an economic system, and civilization. Simultaneously, the future metaverse should be created by users, while Roblox only provides them with tools and technology.

Matthew Ball, a VC analyst, sees the metaverse as more than simply a virtual space, a virtual economy, a game, an app store, or a UGC platform; instead, he views it as a persistent, stable, and real-time existence that can accommodate many participants, spans across the virtual and the real world, and owns closed-loop economic systems and data, asset interoperability, and users who continue to produce content.

Ma Huateng, CEO of Tencent, believes extended reality (XR) is a process from quantitative change to qualitative change, which means the integration of online and offline, and physical and electronic means. The portal between the virtual and the real world is open. It is committed to helping users achieve a more realistic experience. From the consumer Internet to the industrial Internet, application scenarios have also been opened. Communication and social networking are becoming video-based, video conferencing and live streaming are on the rise, and games are becoming cloud-based. With new technologies such as VR and new hardware and software being promoted, he is sure that a major change is coming.

Xu Siyan, a senior researcher at the Tencent Research Institute, views the metaverse as a continuous virtual space that can be shared. In this virtual space, people can not only entertain, but also socialize and make purchases, etc. None of these behaviors requires human participation, but can interact with reality. In an ideal metaverse, players can do almost everything in the real world in a virtual space.

Cai Haoyu, CEO of miHoYo, hopes that in the next 10 to 30 years, we can create a virtual world like those in *The Matrix* and *Ready Player One*, where billions of people around the world can live.

Having summarized their understanding of the metaverse, it is notable that there is a certain commonality among the aforementioned experts: it is a super-large digital community generating the ultimate immersive experience, a rich content ecology, a super-space social system, and a virtual-real interaction economic system. This can reflect real human society.

Myself: the so-called metaverse is a product driven by a variety of technologies, namely, a world where virtuality and reality are mixed. Both the individual and the physical world will become ubiquitous and accessible thanks to this technology. The unknown boundary or blurred line between the real and the virtual worlds is what we call the metaverse.

Several Core Elements

(1) A stable economic system

The most "sci-fi" thing about the metaverse is that it may create a real socio-economic system. In the present game economy, plenty of gamers spend time collecting digital assets to sell them inside or outside of the game. This labor is usually short, repetitive, and limited to a handful of applications, but the variety and value thereof grow as the metaverse evolves itself. Just as painters can sell and monetize their works in the form of NFT tokens, players' virtual items and creations in the metaverse can be converted into digital assets too. Games are more than games once digital assets are introduced into the metaverse. Individuals and enterprises can create, own, invest, and sell assets. Their work creates value.

Like reality, the metaverse requires an independent economic system and independent economic features. In fact, the core of the metaverse lies in the credible bearing of people's asset rights and social identities. This replication of the underlying logic of the real world makes it a robust platform, where any user can create, and the fruits of labor are guaranteed. The content of the metaverse is interoperable and virtual assets created by users can be circulated without platform constraints. This affords it economic and cultural prosperity similar to real life, while it forms a relationship with the real economy.

Based on this, there is no discernable difference between people's labor creation, production, and transaction in the metaverse and real life. Everyone can create, trade, and "work" for rewards, and users' virtual rights are guaranteed. When digital assets are introduced into the metaverse, players' virtual items and creations can be converted into digital assets too. Individuals and enterprises can create, own, invest, and sell assets; their work creates value. For example, a virtual house that one user builds in the metaverse can be easily traded without platform restrictions, and exchanged for other items in the metaverse or the real world, at a market-determined price.

(2) STRONG SOCIALITY

When it comes to sociality, Maslow's pyramid of human need levels is inevitably involved. The generation of social relationships is closely related to the stage of life. In reality, people have social relationships the moment they are born, which continue to expand into a web. Since birth, we have social needs, which are not entirely of our own initiative. The purpose of our social behaviors at this stage is more about obtaining basic guarantees such as healthy growth, personal safety, education, and financial income. For example, the relationship between us and our guardians or first teachers; the desire to socialize begins to manifest when self-consciousness emerges. As the mind expands with age, such desire drives us to build more interesting social relationships.

Before the advent of the Internet, people used to be limited to a small social circle, including (in most cases) a strong acquaintance from birth – family members; then there will be the necessary social connections during social survival and development, such as those with classmates and teachers at school, and later with colleagues at work.

The Internet has quietly brought a brand-new human social organization and survival mode into our society. It has assembled a gigantic group beyond the earth's space – the netizens. In the 21st century, new forms and characteristics of society are surfacing, and individuals in the age of network globalization are converging into new social groups. The emergence of the Internet has expanded our various circles indefinitely. Weibo has narrowed the distance between common folks and celebrities; Zhihu has gathered all kinds of novel questions and answers in the world; Momo has provided us a chance to bond with partners nearby.

The metaverse will give us a more colorful online social scene than the Internet. People can attain emotional sustenance as well as a sense of belonging. There, users can experience different content, make friends, create their own works, and engage in social activities such as trade, education, and meetings.

(3) Immersive experience

Immersive experience is deeply connected to human evolution. As early as ancient Greece, scholars such as Plato have described the characteristics of sensory experience. Later, in the 19th century, German philosopher Friedrich Nietzsche – inspired by Heraclitus and other classic thinkers – pointed out, that games are never random play, but are extremely focused creations, so that endogenously an order is formed. On the one hand, players are totally engaged; on the other hand, they get to observe in silence, which is the secret of their great pleasure from the games. Put another way, the process people go through to obtain immersive experiences is not only a tireless process of construction and exploration, but also to obtain great pleasure and beauty.

Immersive experience advances as productivity progresses. Before the industrial society, due to the limitations of technological equipment and consumption levels, the immersive experience people obtained used to be fragmented and accidental; therefore, it was difficult to become a consumption form that people could widely pursue. Since post-industrialization, human consumption has surpassed all conventional stages in both price and quality, value for money, and full enjoyment. The application of new audio-visual technology, AI, 5G, AR, and VR has made it practically feasible to develop a high-value consumption form. Thereafter, vigorous human development and the extensive pursuit of experience have been promoted. The metaverse is an immersive, high-experience, and gamified lifestyle.

B. Joseph Pine, an American scholar, defines experience as the fourth kind of economic provision in human history in the *Experience Economy*. The agricultural economy provides natural products, the industrial economy offers standardized goods, the service economy renders customized services, and the experience economy gives personalized experiences. When standardized products, commodities, and services begin to exhibit excess capacity, only experience is a high-value carrier that is in short supply.

At present, the immersive experience of haunted houses, escape rooms, and dramas is widely popular among young people. In *Sleepless Night*, an immersive drama masterpiece, the audiences are no longer bystanders sitting below the stage, but are part of the drama. Various explorations wait are triggered by the audience, and there are real props such as letters that can be opened, tea cups that hold real beverages, wet bloodstains, and cold winds.

An immersive experience is characterized by grand spectacles, super shocks, full experiences, and strong logic. Among them, the grand spectacle integrates advanced technologies such as new audio-visual presentation, AI, high simulation, mixed reality, and human-computer interaction. It enters a vast field beyond the reach of human eyes and ears, including micro and macro fields, transcending the experience the audience can have in nature and daily life. This super-intensive expression is an unforgettable experience, mobilizing the audience's vision, hearing, and smell greatly. From the superficial sensory to the deep philosophical experience, people enter the pre-designed scenarios, creating a more realistic audio-visual world that can operate autonomously, though it is purely artificial.

The metaverse should also be an alternative to the real world. It is designed according to modern logic and effectively controlled by intelligent means, gathering highly integrated forms of various experiences. Under the general trend of mixed virtuality and reality, information terminals develop along two threads of high-frequency interaction and simulation. The breakthrough of XR equipment based on VR and AR in the fidelity of simulation will bring a qualitative improvement to the immersive experience.

(4) Rich content

At present, there is no unified standard for digital content in the world, however, it is basically accepted both in China and abroad that there are eight branches in the digital content industry: digital games, such as home console, computer, online, arcade, and handheld consoles games; computer animation, as applied in entertainment including films, tv series, games, and networks, and industrial and commercial applications like architecture and industrial design; mobile content, such as text messages, ringtone downloads, news and other data services; digital audio-visual application, such as the digitization of traditional movies, TV series, and music, digital karaoke; digital learning, such as remote education, educational software, and various course services; network services, including an assortment of ICP, ASP, ISP, IDC, MDC; digital publishing, such as digital publication, digital libraries, various databases, and so on; and content software, which mainly refers to the provision of application software and platforms required for digital content services.

The production process of the digital content industry means that it cannot develop smoothly without efficient content transmission channels and solid IT support. Also, it cannot do without a high level of content originality, such as education, entertainment, consulting, art and culture. That is to say, digital content requires the support of an industrial cluster to fully function. Thus, industry clusters come into being, covering a wide range of industries, including content generation, processing, and services.

Should the metaverse want to serve as a long-term virtual space for users to live in, it must develop content tools and an ecology, open third-party interfaces to lower creation thresholds, and establish a self-evolution mechanism with the help of AI. This includes efficient content transmission channels and solid information technology, openness and freedom of creation, and a sustainable environment for content generation.

1.3   The Ultimate Internet Form

While the metaverse appears as a series of real-time and ultimately interconnected online experiences, it is actually empowered and characterized by transformative trends long known to people deeply familiar with cutting-edge technologies and the industry, including shared social spaces, digital payments, gamification, and more. And from a completely physical world to a virtual world intertwined with it, and then to a completely virtual metaverse – in which the Internet is the biggest technical variable. Ultimately, the metaverse represents the evolution of the third-generation Internet.

From Birth to Commercial Use

On October 4, 1957, the Soviet Union launched the first man-made satellite, known as Sputnik, which opened the door to the vigorous pursuit of interconnection technology during the Cold War by the military and political authorities of the United States.

President Eisenhower granted funds to establish the Advanced Research Projects Agency (ARPA) in 1958, with the purpose of centralizing control of all advanced military research projects and preventing vicious competition within the military at all levels. Over the next few years, as ARPA continued to expand, it gradually widened its research scope. Regarding command and control, the research field of human-machine interaction gained prominence as time-sharing computing systems began to be widely applied on military bases.

In 1962, an interaction theorist, J.C.R. Licklider, took over the control department. With deep faith in time-sharing interactive systems, he began to finance computer research centers of universities and manufacturers, including the Stanford Research Institute led by Douglas Engelbart (where the mouse was invented). Driven by Ricklider and his successors, Ivan Sutherland and Bob Taylor, ARPA funded every major interactive computing development project, including web projects, which were aimed at connecting systems that were incompatible across ARPA sites, and enabling researchers to share computing power and data.

It was not until 1969 that the Internet was truly born from ARPAnet, the predecessor of the Defense Advanced Research Projects Agency (DARPA) of ARPA. In January 1969, Bolt, Beranek, and Newman were commissioned to develop an interface message processor to pave the path for ARPAnet, which was grouping and exchanging messages. Clearly, a set of standards needed to be set up to govern the conversations of computers in a network of interface message processors, the types of applications, and how they work.

Therefore, academics from various DARPA sites requesting interconnection formed the Network Working Group and began applying standards by means of RFC (Request for Comments). After several rounds of revision, they finally reached a consensus: the first result was the birth of Telnet and FTP (File Transfer Protocol). Telnet allows remote users to log in to the system, just like a direct connection through a terminal. FTP solves the problem of file exchange in the network. More importantly, they developed a general system called the NCP (Network Control Protocol), which connects systems symmetrically, instead of a mere client/server (C/S) configuration.

In October 1969, the first two interface message processors were successfully connected. Although one of the computers crashed immediately, the idea proved plausible. Over the next two years, more systems were connected, and the NCP was finally completed in 1971. By then, the US already owned 15 ARPAnet nodes, which were connecting up to 23 systems.

In 1978, Bell Labs proposed UNIX to UNIX Copy Protocol (UUCP). The newsgroup network system developed in 1979 was based on this. Newsgroup (a group discussing a certain topic) was developed in tandem to provide a new way of exchanging information around the world. However, it was not considered part of the Internet because it did not share the TCP/IP protocol. It was connected to UNIX systems all over the world, and many Internet sites took full advantage of it. In a way, the newsgroup was an important part of the development of the network world.

In 1983, when ARPAnet adopted TCP/IP as its basic protocol, other networks had begun to render e-mail, file transfer, and collaboration services outside the ARPAnet architecture. Some of the networks were commercial, such as Systems Network Architecture (SNA) of IBM and Decnet of Digital Equipment Corporation, while others were used for academic research, such as CSNet, Bitnct, and Janet – an English academic network established in 1984.

The most notable was the NSFnet established by the National Science Foundation (NSF). The NSF has established geographically computer wide area networks throughout the US and linked them with supercomputer centers. NFSnet completely replaced ARPAnet in June 1990, becoming the main Internet network, and gradually it has evolved into today's Internet. Arguably, its greatest contribution was making the Internet accessible to society as a whole, rather than keeping it exclusive to computer researchers and government agencies as before.

Web 1.0

Although moderners find it unimaginable to live without the Internet, in fact, the modern Internet was born merely 43 years ago. And the best-known form of it – the World Wide Web – is only in its 30s.

In the early 1990s, driven by a series of infrastructure and application software developments, networks began to change. By 1990, the Internet had connected 300,000 hosts and over 1,000 newsgroups using the Usenet protocol. The original intention of programming Usenet was to connect to Unix hosts via built-in UUCP protocol, but Usenet was eventually incorporated into TCP/IP. And ARPAnet was officially retired while only the Internet continued to grow. In 1991, the release of the World Wide Web greatly changed the face of the Internet, pushing it into Web 1.0.

In 1994, Netscape released the first browser in history. Though it was not as diverse as the browsers of today, it truly opened up a new way to display information. In the same year, the founder of Css released css, enabling the web pages to be beautified, and styles gradually became more varied. Also, the famous W3C Consortium, the World Wide Web Consortium, was founded. Now, it formulates all technical web standards in the world.

In 1995, JavaScript, developed by Netscape, came out. It has made the browsers intelligent, letting them operate on their own. For example, at login and registration, they can judge whether the data entered by the user is correct and when to pop out an error prompt, and encrypt and protect the data entered. The year JavaScript was born is of historic significance because it created infinite possibilities in front-end development.

In 1996, Microsoft launched the iframe, which frees the browsers from synchronous rendering only. Instead, webpages are no longer loaded and rendered in sequence, but asynchronously, which has greatly improved the speed of opening webpages. In the same year, the W3C held its first meeting, where the first specification versions of css and html were released. This fundamentally solves the chaos of non-uniform versions, and enables everyone to formulate norms together and connect.

In the era of Web 1.0, technically, the front-end Web was born, and the future development path was laid. Commercially, the Internet moved from being for exclusive use in a certain field into public use for general society. The Web 1.0 era is one of the multi-horse races, portal websites, and of the most outstanding content characteristics.

The nature of Web 1.0 is to aggregate, unite, and search. The aggregated objects are massive and jumbled network information. This is the smallest independent content data created by people in the era of web pages, such as a blog article, a comment on Amazon, or a modification of one Wikipedia entry. It can be as small as a short sentence or as large as a few hundred words. It can be an audio or video file, or every click of support or objection from passing users. In fact, at the beginning of the Internet, its core competitiveness in commercialization lay in the effective aggregation and use of these tiny contents. And strong search engines such as Google and Baidu gathered these trivial and scattered values, forming a powerful and effective function.

Netscape, Yahoo, and Google, among others, have made great contributions to Web 1.0. Netscape developed the first widely-used commercial browser, Jerry Yang of Yahoo proposed the yellow pages of the Internet, and Google launched the widely popular search service. As for profitability, Web 1.0 is based on one common point – a huge number of clicks. Whether it is early financing or subsequent profiting, it relies on a large number of users and click rates. Listing or developing value-added services on the basis of click rates, and using audiences to determine the level and speed of profitability, fully reflects the eyeball economy of the Internet.

Notably, despite the potential profitability of Web 1.0, it still does not have a good business model: the idea of product management is unpopular, and the concept of product operation remains only budding. Moreover, Web 1.0 addresses people's needs for information search and aggregation only, but not their needs for communication, interaction, and participation. Moreover, it is static, poorly interactive, slowly accessible, and the interconnection between users is rather limited.

Web 2.0

In March 2004, during a brainstorm at O'Reilly Media, Web 2.0 was explicitly proposed. Subsequently, under the vigorous promotion of O'Reilly Media, the world's first Web 2.0 conference was convened in San Francisco in October that year. Thereafter, the concept of Web 2.0 swept the globe.

At present, the more classic definition is given by Blogger Don in his Interpretation of Web 2.0. "Represented by websites like Craigslist, LinkedIn, Tribes, Ryze, Friendster, Del.icio.us, 3Things.com, and so on, and centered on the application of social software such as blog, TAG, SNS, RSS, and Wiki, Web 2.0 is a new model of the Internet, based on new theories and technologies such as six degrees of separation, xml, and ajax. It is a general term for a new class of Internet applications in contrast to Web 1.0, and a revolution from core content to external applications."

While Web 1.0 is believed to have met people's needs for information, Web 2.0 mainly caters to communication, exchange, participation, and interaction needs. Though Web 2.0 also emphasizes content production, it has been expanded from professional websites to individuals, from the institutionalized and organizationally controlled production to more self-organized, random, and self-checked production. There is a decentralization trend for such content production. And the purpose of individual content production often lies not in the content itself, but in extending human relationships in the network society by using the content as a link and a medium. Society is no longer a "mimic society," but part of real life.

The blog is a typical embodiment of Web 2.0. A blog is an easy-to-use website where users can freely publish information, communicate, and engage in other activities. It lets individuals express themselves on the Web and receive feedback from like-minded people. Bloggers are both archive creators and archive managers. The emergence of blogs ignited a revolution in the online world, greatly lowering the technical and financial thresholds for website creation and thus enabling every Internet user to build their own online space easily and quickly. Therefore, their need to transform from a simple information recipient to an information supplier is satisfied. It is from blogs that the popular Weibo (literally meaning micro-blog) was developed.

The blog has succeeded in building an ecosystem in the Web 2.0 era, which contains three levels: micro, meso, and macro. Individual bloggers or readers make up the micro level; the crowd absorbed by an individual blog platform is the meso level; the entire blog world is the macro level. The blog ecosystem is no simple supply and demand relationship between "writers" and "readers," or between "performers" and "audience," but an ecological relationship of multiple needs.

Bloggers have a deep psychological motivation to build their self-image, a variety of utilitarian appeals, and an external pursuit for social reward. Blog audiences are mainly driven by a sense of social belonging and they also have various external goals. The appeal of both bloggers and blog audiences echoes and serves each other, which constitutes the ecology of the blog world.

The macro system of the blog world also produces its unique cultures, communities, customs, systems, and even institutions, which are not water without a source, but an inheritance of the relevant factors of the larger ecological system of society. Meanwhile, the particularity of blog ecology will give them some distinguishing characteristics, such as relative freedom, openness, tolerance, and variability. They act on individuals at the micro-level and affect traditional social-ecological systems through the mediation of the individuals. The interaction of the three blog levels determines the rise and fall of individual blogs and of the entire blog world. More importantly, this interaction is the underlying mechanism by which the blog world casts an impact on society and culture.

Web 3.0 = Metaverse?

From Web 1.0 to Web 2.0, the Internet has revolutionized social life and production. Now, Web 3.0 is coming. While the nature of Web 1.0 was aggregation, union, and search, and that of Web 2.0 was interaction and participation, the nature of Web 3.0 is deeper life participation and life experience.

We may not be able to accurately describe what it will look like, but judging from the front-end signs and development tendencies, Web 3.0 will be a new era of the comprehensive interconnection of objects, accurate expression of objects, accurate human perception, and intelligent interpretation of information. It will be an era of hyperlinks based on the Internet of Everything. There will be an information interaction network that connects the material world and human society in all directions. The resulting network system will be gargantuan, infinitely expanding, rich in layers, and harmoniously operating. We will see a brand new civilization landscape where the real world and the digital world are integrated.

As the growth of any living organism, the Internet is also constantly evolving. Since the official application of the World Wide Web in the 1990s, the Internet has been on a journey where "net" and "people" are constantly approaching each other, merging, and becoming one. From Web 1.0 to Web 3.0, the simulation degree has become higher and higher. Today, virtual life on the Internet is still extending to the depth and breadth of real life so as to achieve a realistic and comprehensive simulation.

Generally, Web 3.0 will be a networked world with a higher degree of virtualization, more freedom, and better manifestation of the personal labor value of netizens. It will be the third world constructed by integrating virtual and physical spaces, and is currently the ultimate form of the metaverse. Conclusively, the metaverse represents all the functions of the third-generation Internet.

1.4 Popularity of the Metaverse

While some people are still confused with the concepts of VR, AR, NFT, and cloud computing, the metaverse – a term that encompasses the above elements while exuding a strong sci-fi vibe – has become a hot topic online. Having attracted the biggest companies to invest generously, the concept of metaverse has become the hottest term in the market. With technology advancing rapidly, games are more involved in people's lives, and with the COVID-19 pandemic propelling the digitization of society, the metaverse is gaining more popularity.

Rapid Technological Advancement

The rapid advancement and application of technologies such as 5G, cloud computing, and VR are driving mankind closer to the metaverse. The improvement of cloud computing promotes cloud game development. 5G will make up for the shortcomings of transmission, drive the integrated development of cloud games, popularize consumer entertainment, and break the time, place, and terminal restrictions on various media Internet services. The application of intelligent technologies in various vertical fields will fully empower media scenarios and improve the efficiency of information production and dissemination.

The continuous iteration of VR/AR technologies and equipment is expected to optimize the digital life experience of users. As more VR/AR equipment continues to be sold and the experience keeps improving, digital services based on VR/AR will penetrate various scenarios and give users a subversive and immersive experience in the metaverse. In fact, a truly immersive parallel world will definitely need the support of VR equipment. Yet in a true metaverse society, there will be screenless holography (screens will be everywhere), and we will be freed from VR devices as much as possible.

Twenty-nine years ago, when the Internet began experiencing rapid development, Neal Stephenson imagined a surreal metaverse world in his science fiction novel *Snow Crash*. People live immersed in the digital world and communicate in virtual avatars. But at the time, VR, the core of the metaverse, was rudimentary and was subject to chip and processing technology.

In 2016, which is widely called the first year of VR, many companies, including Internet giants like Facebook (now known as Meta), entered the game with great enthusiasm. According to CVSource statistics, there were 120 financing events related to VR projects in China that year, reaching nearly 2.5 billion yuan cumulatively. However, VR eventually lost popularity in China and abroad due to the lack of content support.

However, since 2019 the VR product experience has been significantly better, with the network environment upgrade (5G high-speed network effectively reduces the dizziness from delay), the continuous maturity of hardware equipment (the introduction of Fresnel lenses, FastLCD, VR special chips, and so on improves the image definition, reduces equipment weight, optimizes field angles and immersion, etc.), and software product optimization (game developers better know how to develop games based on the characteristics of VR terminals).

The outbreak of the COVID-19 pandemic at the beginning of 2020 has greatly prolonged consumer time at home and expanded demand for domestic recreation. In this situation, VR has once again become a popular form of entertainment. The release of the popular new game *Half-Life* with super immersion further demonstrated the appeal of VR and accelerated its promotion. According to IDC data, global VR shipments in 2020 went up by 2% year-on-year to 5.55 million units – the first time it returned to positive growth since 2017. VR shipment to the US grew 58% year-on-year to 2.84 million units, accounting for about 51% of the world's total. It was a significant increase of 23% when compared with the number in 2016, leading the recovery of global VR demand.

As per statistics from IDC, the size of the global VR/AR market in 2020 was about 90 billion yuan; of this VR made up 62 billion, and AR accounted for 28 billion. The China Academy of Information and Communications Technology predicts the five-year average annual growth rate of the global VR/AR industry from 2020 to 2024 to be about 54% (specifically, VR is predicted to grow 45% and AR about 66%.) It is expected that in 2024, the market size of the two will be close – both reaching 240 billion yuan.

The rapid rise of VR comes not only from the COVID-19 pandemic, but the recent breakthroughs in technical bottlenecks – especially the application of 5G – which greatly improves network transmission and reduces communication delay. After comparing the key performance indicators of 5G and 4G, CCID Consulting, a Beijing-based tech consultancy company, found that 5G reduces delay and further alleviates the dizziness of players. Also, it is possible for relevant host computers to be free from connection cables, or for computing power to be placed on the cloud, so as to make the devices lighter and smaller.

Of course, VR is only one aspect. As underlying hardware, it can bring immediate experience improvement to users. In addition to VR, 5G, cloud computing, and edge computing break the limitations on current computing power and the speed and quality of information transmission. Their wide application will support users in connecting to the virtual world anytime, anywhere; AI based on deep learning improves data collection and processing efficiency, enhances personalized services, enriches the content, and assists in data collection, data processing, and content production in digital life. With the support of such technologies, the degree of digitization of human life will be greater, and the transition to the metaverse faster.

Participation of Games

Roblox, one prototype of the metaverse, is a game platform that provides sandbox game creation and online play. It became an IPO on March 11, 2021, and its stock performance is fairly strong. As the first company to include the metaverse in its prospectus, it has attracted 42 million daily active users and over 7 million content creators. Moreover, it has developed more than 18 million game experiences with more than 22.2 billion hours of player engagement.

According to its prospectus, Roblox had 32.6 million daily active users in 2020, a year-on-year increase of 85%. As of 2021 Q1, its daily active users reached 42 million, a year-on-year increase of 79%. Driven by the lockdown policy during the COVID-19 pandemic, more users were active online. From 2020 Q1 to 2021 Q1, the total time spent online by platform users increased from 4.88 billion to 9.67 billion hours, a year-on-year increase of 98%. In terms of single-user behavior, 2021 Q1 witnessed an increase of 11% year-on-year, and the average every active user spent 2.6 hours per day on the platform. Since then, the time spent online by active users has continued to increase, indicating stronger user stickiness; in 2021 Q1, the average quarterly spending per user reached US $15.50. Since the pandemic, the growth rate of average spending per user has shifted from negative to positive. And new users are quickly adapting to the platform. The active developer ecology, user ecology, and business model of Roblox show the market the future potential of the metaverse, and invite the game industry to successfully enter the arena.

In fact, games naturally have virtual fields and virtual player avatars. Today, the functions of the games have gone beyond the games themselves, and are constantly breaking new ground. In April 2020, the famous American rapper Travis Scott held a virtual concert with an avatar in the game Fortnite. It attracted more than 12 million players around the world, thus breaking the boundary between show business and video games.

During the COVID-19 pandemic, in order to prevent students from missing their graduation ceremony, UC Berkeley rebuilt the campus in the sandbox game Minecraft, and students attended a virtual ceremony in their virtual avatars. ACAI, one of the world's top artificial intelligence academic conferences, held the 2020 symposium on Nintendo's *Animal Crossing: New Horizon*, blurring the boundary between academics and video games. As COVID-19 raged, some parents held birthday parties for their children on Minecraft or Roblox, and many people's daily social interactions became fishing, catching insects, and visiting animals on the island in *Animal Crossing*!

Also, Gucci and Roblox jointly launched The Gucci Garden Experience virtual exhibition. Users could enjoy Gucci design exhibitions on the Roblox platform and purchase several virtual items for a limited time during the exhibition period.

Apparently, a series of immersive experiences based on game kernels have been available in the market, which can be transformed into diversified virtual experiences of real behaviors such as recreation, consumption, and conference. In the future, the boundary between virtuality and reality will continue to be blurred in the metaverse.

The COVID-19 Pandemic Encourages Digitalization

From the perspective of social development, the metaverse will be a real digital society belonging to the next generation as an advanced form of the current Internet and Internet of Things. Since it relies on the establishment of a social system, it must be inextricably linked with the development of the real world. In this context, the COVID-19 pandemic and the social needs of Generation Z will be short- and long-term catalysts, respectively.

On the one hand, the pandemic that swept the world in early 2020 has not yet been brought under control. As of May 17, 2021, the cumulative number of confirmed cases in the world has reached 160 million, and 3.39 million have died. Even with the normalization of pandemic prevention and control, working from home and online business remain the trend. According to Tencent's 2020 financial report, Tencent Conference has become China's largest independent cloud conference APP, with over 100 million users. Taking advantage of the COVID-19 pandemic, it has become widely known.

COVID-19 will also continue to change user habits. In 2020, cloud communications such as Zoom, Microsoft Teams, and Google Meet stood out. Although the market paid slightly less attention to them as the pandemic improved, telecommuting has not disappeared. On the contrary, the payment rate and user growth of Zoom, the leading telecommunication, have greatly exceeded market expectations – employees work more efficiently from home while office costs are lowered. More and more enterprises and individuals have embraced the online space, profoundly changing their production and lifestyle. Some companies, especially those Internet-based, have begun to gradually accept and adapt to this Internet-based virtual telecommuting model.

In addition to the online office, e-commerce, entertainment, medicine, etc., have also become more established online. For example, there are online communities for group purchases, smart logistics and delivery, and fresh food e-commerce. Due to the lower frequency of consumers going out during the pandemic, home delivery demand for fresh food from supermarkets has surged, and the number of active users per day and their hours on major fresh food platforms that offer home delivery have increased significantly. Taking Hema, the supermarket owned by Alibaba, as an example, the Q1 online purchases in 2020 accounted for about 60% of its GMV (gross merchandise value), an annual increase of 10 percent. People have grown more dependent on the Internet to meet their real-life demands.

The introduction of AI and cloud computing has integrated online and offline more closely. Whether it is Uber drivers, Amazon's unmanned delivery, or delivery guys, when they work according to the APPs' instructions, their scheduling models are more online and digital.

The pandemic has also shrunk the space for people's offline activities while they have engaged in more online activities. Time spent on the Internet has witnessed an increase. COVID-19 acted as a catalyst, forcing people to completely disengage from the physical world and reflect on it for the first time; the more time and energy they spend in the virtual world also reinforce their recognition of its value, which paves the way for the arrival of the metaverse on a cultural level.

On the other hand, highly advanced internet technology has brought more and more digital devices. Smartphones, smart watches, tablet PCs, and various other digital devices based on the popularization of smart sensors are increasingly pervasive, taking humanity into an epoch of unprecedented information prosperity. Today, the information that a young man's brain receives differs greatly from that of the past, and the age of first contact with the Internet keeps decreasing. Those born in the 21st century have gradually entered the space of public observation. In contrast to the previous generation, who has just walked into the social age, they are the first generation to grow up in a totally digitalized environment.

Taking China as an example, the whitepaper Chinese Children in the Digital Age published by Wavemaker states that there are as many as 160 million children aged 6–15 in China. The average age at which they start using a computer is 7.8 years, while for a smartphone, it is 7.3 years. Most Chinese kids have access to various devices, video games, and social media before the age of 9. These 160 million Chinese are the first group to have a digital childhood, which is certainly inseparable from technological innovation. With the popularization and promotion of digital technology, digital devices are seen everywhere in life; the penetration of smartphones has reached unprecedented saturation. According to the Ministry of Industry and Information Technology of China, the number of mobile phones owned per 100 people in China has reached 112.2 – higher than one phone per person. And children and teenagers with smartphones are by no means a minority.

Meanwhile, Generation Z is more inclined to express opinions on the Internet and care more about life experiences. The digital life experience provided by the metaverse is another dimension of life – an expansion of human emotions and lifestyles. It is a life that can be restarted, reset, and separated from the physical world. In the metaverse, the sense of experience, achievement, and happiness are all at low cost, and there is no monopoly of resources, which is undoubtedly a huge magnet for younger generations.

Much of the world, including North America, is still faced with the COVID-19 pandemic, and giants such as Microsoft, LinkedIn, and Twitter have extended paid holidays to console their employees. With the rise of e-commerce and live streaming in China, there are more freelance opportunities for Generation Z, and more and more Chinese are becoming full-time or part-time Vloggers. The 2020 Social Responsibility

Report published by ByteDance states that Douyin (TikTok) has created 36.17 million direct and indirect employment opportunities in 2020, and promises that it will support small and medium creators generate a total of more than 80 billion yuan in 2021. Despite the great distance from realizing the world in *Ready Player One*, the trend of social development and change is gradually clearer.

CHAPTER 2

Drive of the Metaverse

The ultimate form of the metaverse is the sum total of many technological innovations.

Steve Jobs once made a famous "necklace" analogy – the iPhone links up single-point technologies such as the multi-touch screen, iOS, high-pixel camera, and large-capacity battery, thus redefining the mobile phones and opening the door to the age of the mobile Internet.

The iPhone is a groundbreaking smartphone with a multi-touch screen; iPhone 3GS kickstarted the 3G era and joined the App Store system; iPhone 4s launched the first voice assistant – Siri, leading to the development of mobile phone voice technology; the iPhone 5 series adopted touch ID for the first time, leading fingerprint identification in the industry; the iPhone 6 series applied Face touch; since the iPhone 8, Face ID has replaced touch ID, and infinity display is widely used.

The true metaverse still requires more technical progress and industrial convergence, but at present, as computing power continues to grow stronger, and technological innovations are brought together, we are approaching the iPhone moment of the metaverse.

2.1 Reconstruction of Computing Power: Building the Metaverse

Computing power is the most important infrastructure for the metaverse. Picture material, blockchain network, and artificial intelligence technology (AIT) that constitute the metaverse can't do without adequate computing power. The metaverse is not an online game, but as an online game, it is a virtual world that hosts activities. Computing power supports the creation and experience of virtual content. More realistic modeling and interaction require stronger computing power.

Supported by computing power, AIT can help users create richer and more realistic content. The proof-of-work (POW) mechanism that relies on computing power is currently the most widely used consensus mechanism in blockchain. Guarding the decentralized value network in the digital world, computing power is an important ladder to the metaverse.

Core of Human Intelligence

The development of human civilization is inseparable from the progress of computing power. Homo sapiens made many calculations thousands of years ago. Subsequently, calculation evolved from tying knots and other activities in tribal societies to abacus in agricultural societies, and to computer calculation in the industrial age.

Computer calculation advanced as computers developed from the relay computer in the 1920s to the vacuum tubes computer in the 1940s, and to the diode, triode, and transistor computers in the 1960s. Of these, the calculation speed of transistor computers is up to hundreds of thousands of times per second. Later, integrated circuits improved it to millions and tens of millions of times per second in the 1980s, and today billions, tens of billions, and hundreds of billions of times per second.

Human biological research shows that there are six brain mantles in the human brain, and the neural connections in them form a geometric progression. The synapses of the human brain beat 200 times per second, while brain nerves can move as much as 1.4×10^{15} times per second, which is the inflection point for computers and AI to surpass the human brain. Notably, the progress of human intelligence is related to the speed of computing tools created by human beings. In this sense, computing power is the core of human intelligence.

In the past, computing power was more regarded as the ability to calculate, but in the age of big data, it has been given new connotations, including the technical capabilities of big data, the ability to instruct problem solving and to systematically calculate programs. On the whole, it can be understood as the ability to process data. William D. Ordhaus, one of the two recipients of the 2018 Nobel Memorial Prize in Economic Sciences, defines computing power in the article "The Progress of Computing" as "the amount of information data that a device can process per second based on changes in its internal state."

Computing power consists of four parts: the system platform, which stores and operates big data; the central system, which coordinates data and business systems, thus directly reflecting governance capability; the scenario, which coordinates cross-departmental cooperation; the data cockpit, which directly reflects data governance and application ability. Conclusively, computing power is the integration and accumulative application of multiple functions.

When it is used to solve practical problems, computing power alters the existing production mode and enhances decision-making and information screening of the existent. Meanwhile, diversified scenario applications and continuous iteration of new computing technologies have freed computing and computing power from the limitations of data centers, expanding them to the cloud, network, edge, and terminal scenarios. Computing begins to transcend tool and physical properties, evolving into a new form.

From a functional perspective, as humans continue to demand more from computing, it has gradually formed perception, natural language processing, and thinking and judgment capabilities based on a single physical tool. With the help of a series of digital software and hardware infrastructures in the form of technology and product, it has penetrated into all aspects of social production and life rapidly.

Computing is needed for several purposes, performing multiple functions today. Having undergone a complete transformation from "old" to "new," it has become an extension of human capability. As the American architect, Negroponte wrote in the preface of *Being Digital*, "computation is no longer just about computers, but also determines our being." Computing power is becoming a crucial factor in the social lifestyle of millions.

Critical Infrastructure

Computing power is the most important infrastructure for building the metaverse. Picture material, blockchain network, and AIT simply can't do without its support.

The graphic display of the virtual world is inseparable from the support of computing power. Computer graphics renders model data to every pixel in the entire image according to the corresponding process; thus a huge amount of computation is required. The 3D effect displayed on the current user's device is usually composed of polygons. Whether it is the interaction of the application scene, the various games, or the fine 3D models, most of the models inside are created via polygon modeling.

The movements, actions, and character changes are due to different lighting, rendered in real time by the computers based on various calculations. The rendering process includes five steps: vertex processing, primitive handling, rasterization, fragment processing, and pixel transfer operation. Each step needs the support of computing power.

Computing power supports the creation and experience of virtual content in the metaverse; more realistic modeling and interaction require stronger computing power as a premise. Moreover, the flywheel effect of game development and graphics card advancement constitutes hardware and software foundation for the metaverse. From the perspective of the gaming industry, every major leap comes from the continuous progression of computing power and video processing technology.

3A (a lot of time, a lot of resources, a lot of money) games often take high-quality graphics as the core selling point. They make full use of or even overexploit the performance of the graphics card, which triggers a "graphics card crisis." When game consumers pursue high quality and high experience, they will inevitably need devices with strong computing power, thus triggering the flywheel effect of game development and graphics cards advancement. This has already taken place in popular games like *Need for Speed.*

One of the biggest challenges in building the metaverse is creating sufficient high-quality content, especially as production costs are prohibitively expensive. 3A games often require several years of input from a team of hundreds of members, and the UGC platforms are exposed to risk in quality assurance. To that end, the next big development in content creation will be the shift to AI-assisted human creation.

Though today only a handful of people can become creators, this AI-complementary model will completely democratize content creation. Everyone has the chance to be a creator with the help of AI: AI tools can translate advanced instructions into productive results, taking care of the "heavy lifting" of coding, drawing, animation, etc. In addition to the creative phase, there will also be NPCs involved in social activities within the metaverse. They will have their own communication and decision-making capabilities to further enrich the digital world.

POW that relies on computing power is currently the most widely used consensus mechanism in blockchain, and a decentralized value network requires safeguarding computing power. The POW mechanism is a proof-of-work mechanism, which means the contention for bookkeeping rights (also the contention for token economic incentives) is the contention that determines the victory or defeat via computing power. This is also the least wasteful situation from an economic perspective. In order to maintain the credibility and security of the network, it is necessary to supervise and punish malicious nodes, prevent 51% attacks, etc., all of which are carried out under the constraints of the POW consensus mechanism.

Promote Computing Power Development

The metaverse places extremely high demands on computing power. As the most critical infrastructure of the metaverse, it has greatly changed the face of society. It should be noted that the current computing power architecture still fails to meet metaverse demands for low threshold and high experience. However, the advancement of edge computing, quantum computing and chip architecture will improve computing power and clear the obstacles for metaverse development.

(1) EDGE COMPUTING

Usually, there is one-time and valuable data amongst a sea of data – the types of data are in a mess. To sort and filter through it, computer arithmetic is necessary. Owing to the constraints on local computer power, more and more applications rely on cloud computing; thus demand for it is slowly expanding. Of course, while cloud computing renders services, the optimization of computing power systems is progressing too.

However, cloud computing, despite its power, has its limitations. Generally, when data is processed, if only cloud computing is used for data processing, there will inevitably be latency. Judging from the entire process, all data is first transmitted to the central computer room through the network, then processed through cloud computing; after processing is completed, the result is returned to the corresponding locations. For such data processing, there are two more obvious problems.

One is the timeliness of computing power. Data feedback will be delayed, mainly due to mass data transmission. The transmission will be blocked by limited bandwidth resources, which will prolong the response time. The other problem is the effectiveness of computing power. All data will be transmitted to the central computer room, but some are useless, and due to the lack of preprocessing, this will waste some cloud computing power.

The "center-edge-end" operation mode solves the predicament of cloud computing punctually. It has been fully applied in the telecommunications network age, and to a certain extent, ensures the orderly and effective operation of the entire network. Specifically, the center refers to the stored-program-controlled (SPC) exchange center, the edge refers to the SPC exchange, and the terminal to the telephone.

In the age of the Internet, the "center-edge-end" mode continues. Data Center-CDN-Mobile Phone/PC is its application in the Internet age. Among them, CDN (Content Delivery Network) is designed to avoid network congestion as much as possible, provide customers nearby with the required content, and improve the response speed of website browsers. This marginalized design optimizes the distribution or delivery of online content, thereby improving network efficiency and user experience.

However, traditional CDN has limitations, too. It focuses on caching, which is obviously not enough for today's cloud computing + Internet of Things (IoT). In such times, there is an explosion of data – the data transmitted increases geometrically, which greatly tests the entire network capacity.

From the perspective of the traditional CDN operation mode, the data generated by the terminals will have to be returned to the central cloud for processing. In the case of massive data transmission, there will be two obvious problems: use-cost and technology implementation. From the perspective of use-cost, the usage rate of traditional CDN has always been high (mainly because the charging of fees is not flexible enough to be on-demand). And technical problems are manifested in bandwidth. Taking the mobile network as an example, the traditional CDN system is not usually deployed in the provincial IDC room, but inside the mobile network. Therefore, data must travel through a long transmission path to reach the data center.

In order to improve the timeliness and effectiveness of data processing, edge computing arrived at the right moment. The edge represents the edge node. As the name suggests, edge computing refers to end-to-end computing services for the central platform at the near-end or data source. The concept of edge computing is similar to that of an octopus: the nodes can be understood as the tentacles, which are a type of distributed computing.

Edging computing is most characterized by rendering services on the network edge closer to the terminals. Such a design can meet the key needs of various industries in terms of digital agile connection, real-time business, data optimization, application intelligence, and security and privacy protection. Its advantages promote intelligentization and connect the physical and digital worlds.

As another new model after distributed computing, grid computing, and cloud computing, edge computing takes cloud computing as the core, the modern communication network as the approach, and massive, intelligent terminals as the forward position. It integrates all three and is believed to be an important solution to future digital problems.

Certainly, the acceptance and development of edge computing will take some time. John Vicente, CTO of Stratus Technologies, divides the maturity of edge computing into four levels: from an isolated static system 1.0 to invisible, adaptive, and self-managing systems 4.0.

Edge Computing 1.0 is about how to secure, manage and connect machines and devices to activate the digital edge. This stage contains only the basic competencies required to successfully operate a business in a digital world.

In Edge Computing 2.0, it begins to adopt open, software-defined technologies. Software-defined technologies extract various functions from the underlying computer hardware and enable them to be executed in software.

For example, with software-defined networking (SDN), enterprises can more easily manage the network by modifying various attributes – including routing tables, configurations, and policies – from a centralized control platform rather than modifying the attributes of each switch one by one. Similarly, software-defined technologies have enabled cloud-based security services, thus freeing enterprises from the necessity to run firewalls and intrusion detection/prevention systems themselves.

In Edge Computing 3.0, IT and OT will be truly integrated to possess a series of elastic and real-time capabilities. Today, there are still many industrial areas that IT has not laid a finger on. For example, factories require mechanical control systems to perform deterministic behavior and ensure safety. These control systems were born in the field of operational technology instead of IT.

Ensuring the capabilities of Edge Computing 3.0 is necessary for a successful transition to Edge Computing 4.0. In the 4.0 stage, IT and OT infrastructure and operations will be mixed with AI, so that a self-managing, self-healing, and automated industrial field will be born. Once the machine malfunctions, the AI system can diagnose and fix it – without human intervention. Edge computing is inevitable and will strengthen computing power. It can provide high-quality interactive experiences for all devices.

(2) FROM GPU TO DPU

For a long time, computing power was divided equally between the central processing unit (CPU) and the graphics processing unit (GPU). It's also because CPU and GPU can power new super-large data centers, which free computation from the cumbersome constraints of PCs and servers.

Having been at the heart of every computer or smart device since the 1950s, the CPU is the only programmable element in most computers. Moreover, since its invention, engineers have never stopped trying to make the CPU achieve the fastest computing speed with the least energy consumption. Even so, the CPU is too slow to conduct graphics calculations – in this context, the GPU was invented.

Nvidia came up with the concept of GPU, giving it the lofty status of an independent computing unit. The GPU is a specialized circuit that quickly manipulates and modifies memory in buffers. It is widely used in embedded systems, mobile devices, PCs, and workstations because it can speed up the creation and rendering of images. Since the 1990s, the GPU has gradually become the center of computing.

In fact, the first GPU was only used for powerful real-time graphics processing. Later, with its excellent parallel processing capability, it became the optimum choice for various accelerated computing tasks. With the development of machine learning and big data, many companies use GPU to accelerate the execution of training tasks, which is also common in present data centers.

Most CPUs are not only expected to complete tasks faster and more efficiently, but are required to quickly switch between different tasks to ensure instantaneity. It is precisely because of such requirements that the CPU often executes tasks serially. The GPU is designed to be completely different, however. It is expected to increase the handling capacity of the system.

The difference in design philosophy is ultimately manifested in their respective number of cores. Often, the GPU tends to have a higher core count. Certainly, the differences between CPUs and GPUs complement each other nicely, and their combination has powered new super-large data centers over the past few decades.

In recent years, existing general-purpose CPUs and GPUs have been unable to fully meet rapidly changing application requirements, since the CPUs are subjected to more and more network and storage workloads. According to IDC statistics, the growth of global computing power has significantly lagged behind that of data in the past decade. The global demand for computing power doubles every 3.5 months, far exceeding its current growth rate.

Driven by this, the global computing, storage, and network infrastructure are experiencing a fundamental change. Some complex workloads cannot be handled well on general-purpose CPUs. In other words, the CPU-centric data center architecture can no longer meet the demands.

Song Qingchun, senior director of market development in the Asia Pacific region of Nvidia commented: "In the past, the computing scale and data volume were not as large, and the von Neumann architecture solved the problem of computing performance improvement. As the data volume expands, and the development of AI, the traditional computing model will cause network congestion, and it will be challenging to continue to improve the performance of the data center."

The emergence of the Data Processing Unit (DPU) may resolve this dilemma. As a major category of newly developed special-purpose processors, DPU provides computing engines for high-bandwidth, low-latency, and data-intensive computing scenarios. At present, the DPU has become one of the three pillars of the data-centric accelerated computing model, and it will also become the offload engine of the CPU, releasing the computing power of the CPU to the upper layer.

As per the chronological order and characteristics of the technology, the development of the DPU is divided into three stages:

The first is the smart device stage, a.k.a. the prehistoric times of the DPU. At this stage, the easiest way to solve the traffic problem between nodes is to increase the processing capacity of the network card. By introducing SoC or FPGA on the network card, some specific traffic applications are accelerated, thereby enhancing network reliability and reducing latency.

Xilinx and Mellanox started relatively early in this field. Unfortunately, due to insufficient strategic capabilities, they missed the opportunity for further development, were gradually replaced by DPU, and eventually fell away. Mellanox was acquired by Nvidia and Xilinx by AMD. Therefore, intelligent ethernet cards exist as the application product of the DPU.

The second stage is the data processing chip stage. It was first proposed by Fungible in 2019, but it did not attract much attention. After Nvidia repackaged the acquired Mellanox, it redefined the concept of the DPU in October 2020. And the redefinition made it a hit.

Specifically, the DPU is redefined as a new type of programmable processor that integrates three key elements: an industry-standard, high-performance, and software-programmable multi-core CPU that is usually based on the widely used Arm architecture; a high-performance network interface that analyzes, processes, and efficiently transfers data to GPUs and CPUs at high speed; a variety of flexible and programmable acceleration engines that can offload applications such as AI, machine learning, security, telecom, and storage, and improve performance.

The third stage is infrastructure chips. It was proposed by Intel, and it became the FPGA+Xeon-D model, which was placed on an intelligent ethernet card by means of a PCB version. It is clearly observed that Intel positions the IPU as a small "plug-in" CPU on the host CPU. Moreover, in the future, this "plug-in" CPU and FPGA will be packaged into one chip to form a dual-CPU system interconnected through the PCIe bus.

Indeed, regardless of the stages, all DPU capabilities are important for enabling secure, bare-performance, cloud-native, next-generation large-scale computing. As Nvidia CEO Jen-Hsun Huang stated in his speech, "it will become one of the three pillars of future computing … the CPU is used for general-purpose computing, the GPU for accelerated computing, and the DPU that transmits data in the data center for data processing."

On the one hand, the GPU is more secure. Because the control plane can be separated from the data plane within the system and between system clusters, the DPU can also perform tasks such as networking, storage, and security that used to require CPU processing. This means that if the DPU is used in the data center, a lot of its computing power can be released to execute a wide range of enterprise applications.

On the other hand, the DPU also frees up server capacity so that it can return to application computing. In some systems with great I/O and heavy virtualization, the kernel cost is halved, so the handling capacity is two times higher. In addition to the kernel cost, the cost of the entire machine is calculated, including its memory and I/O and the amount of work freed.

In addition, the DPU's rich, flexible and programmable acceleration engine improves AI performance and machine learning applications. All of these DPU functions are critical to achieving isolated bare-computer cloud-native computing. It is foreseeable that an integrated architecture from the CPU to the GPU, and to the

DPU will make management procedures and schedulers easier. From the edge to the core data center, unified architecture, management, and scheduling may come true in the near future.

(3) QUANTUM COMPUTING

The development of quantum computing will further transform computing power. It is a new computing mode that follows the laws of quantum mechanics to control quantum information units for calculation, which differs completely from the computing modes in use.

When understanding the concept of quantum computing, it is often compared to classical computing. In classical computers, the basic unit of information is the bit. Everything these computers do can be broken down into 0s and 1s, and simple operations of 0s and 1s. Similar to the way that traditional computers are made of bits, quantum computers are made of quantum bits, or qubits. One qubit corresponds to one state. However, the state of a bit is 0 or 1, while a qubit is a vector. More specifically, the state of a qubit is a vector in a two-dimensional vector space, called the state space.

Classical computing performs calculations digitally using binary, which is always in a definite state of 0 or 1, while quantum computing can realize the superposition of computing states with the help of the superposition property of quantum mechanics – there is not only 0 and 1, but a superposition in which 0 and 1 coexist.

The 2-bit register in a common computer can only store one binary number (one of 00, 01, 10, 11) at a time, while a 2-qubit register in a quantum computer can hold a superposition of all 4 states simultaneously. When the number of qubits is n, a quantum processor performing one operation on n qubits is equivalent to performing 2n operations on classical bits.

In addition, due to the characteristics of quantum entanglement, quantum computers theoretically have faster processing speeds and stronger processing power on some specific problems than classical computers that use the strongest algorithms.

In recent years, quantum computing technology and industry have exhibited accelerated development. The breakthroughs in quantum computing technology are mostly related to three factors: the length of time that qubits can maintain their quantum states; the number of qubits connected together in a quantum system; and the grasp of what malfunctions in quantum systems.

The length of time that a qubit can maintain a quantum state is called the qubit coherence time. The longer it maintains a "superposition" (qubits represent both 1s and 0s), the more program steps it can handle and the more complex calculations it can perform. IBM took the lead in introducing quantum technology into practical computing systems, increasing the coherence time of qubits to 100 microseconds. When qubit coherence time reaches the millisecond level, it can support a computer in solving problems that today's classical computers cannot.

Following the breakthrough in the number of qubits connected together in quantum systems, Google announced the use of the 54-qubit processor Sycamore in *Nature* in October 2019. It has achieved quantum superiority. Specifically, Sycamore can complete the specified operation in 200 seconds, while the same amount of computation would take 10,000 years to complete on Summit, the world's largest supercomputer. This was the first time in human history that quantum supremacy was achieved in an experimental environment, and *Nature* labeled it as a milestone in quantum computing history.

One year later, the Chinese team announced the birth of the quantum computer Jiuzhang, challenging Google's quantum supremacy and leading the world's computing power. As a quantum computer with 76 photons and 100 modes, Jiuzhang can perform Gaussian boson sampling one hundred trillion times faster than the current fastest supercomputer, Fugaku. For the first time in history, a quantum computer built using photons outperformed the fastest classical supercomputer.

Quantum mechanics is a branch of physics that studies the behavior of subatomic particles. Using advanced quantum mechanics, quantum computers – which break through the limits of classical Newtonian physics – make it possible to achieve exponential growth in computing power.

For example, potential applications of quantum algorithms for AI include quantum neural networks, natural language processing, traffic optimization, and image processing. Among them, quantum neural networks, as a research field formed by the overlap of quantum science, information science, and cognitive science, can utilize quantum computing power to improve the information processing capability of neural computing.

As for natural language processing, in April 2020, Cambridge Quantum Computing announced the success of a natural language processing test performed on a quantum computer. It was the first successful validation of quantum natural language processing application on a global scale. Researchers used the "eigen quantum" structure of natural language to translate sentences with grammar into quantum circuits. They implemented the process of program processing on quantum computers and received answers to the questions in the sentences. With quantum computing, it is expected to achieve further breakthroughs in "semantic perception" in natural language processing.

Times are changing: computing power enables the underlying logic of metaverse technology system, and its impact on people and the world has been embedded in all aspects of social life. The future of the metaverse created by computing power will be one where everyone can benefit – one that is completely different from the past. And it is imperative that we base ourselves on computing power and develop it.

As technologies advance, computing power will be detached from the computing mode in centralized computer rooms of the past to multi-dimensional real-time computing processing modes, such as front-end equipment, edge computing, and cloud computing.

2.2 5G: Network Base of the Metaverse

Throughout the history of communication development, transmission rate improvement has always been the focal point. To realize massive real-time information exchange and immersive experience in the metaverse, it requires ongoing improvement in communication technology and computing power. Consequently, users will experience little latency and high simulation fidelity, which was difficult to reach in the 4G era.

The arrival of 5G has provided a great chance for application innovation. With the upgrade of communication, the foundation of the metaverse is extremely robust.

The Popularity of 5G

In fact, as early as June 2015, ITU formally proposed the concept of 5G at the 22nd meeting of ITU-R WP5D. 5G not only outperforms previous generations of mobile communication systems in key capabilities such as user experience, connection density, end-to-end latency, peak speed, and mobility, but provides technical support to realize massive device interconnection and differentiated service scenarios. The commercialization of 5G in 2019 was the official arrival of the 5G era.

As of today, the global industry has reached a consensus on the concept and key technologies of 5G. It will further enhance the consumer experience on mobile broadband applications. Driven by innovation, it strives to become a software-based, service-based, and agile network that serves smart homes, smart buildings, smart cities, cloud work, cloud entertainment, industrial automation, auto-pilot vehicles, and other vertical industries.

Without a doubt, 5G has become a headline topic in contemporary global mobile communications. As critical information technology of the information age, it also plays an important role in every state's digital construction. In the past couple of years,

China, South Korea, and the US – as the first countries to use 5G commercially in the world – have exhibited different characteristics in 5G development. They have provided a valuable experience for the subsequent deployment of 5G applications.

China has always regarded 5G as a major strategic opportunity to technically support domestic enterprises in conducting basic technical studies. On the national strategic level, The Fourteenth Five-Year Plan for National Economic and Social Development and the Outline of Vision 2035 mentioned 5G construction and application three times; the 5G Application "Set Sail" Action Plan (2021–2023) released this year proposed eight special actions and four key projects to illuminate the path for the innovation and development of 5G applications in China in the upcoming three years.

In terms of industry applications, as of the end of May 2021, there have been more than 819,000 5G base stations in China, and the number of 5G terminal connections increased to over 335 million. The average download speed of 5G that users experience in China is 374.2 Mbps, and the upload speed has reached 31.4 Mbps – both are over tenfold that of 4G.

In terms of technological innovation, thanks to the solid technical foundation of mobile communications, the Chinese 5G industry has matured from standard freeze to commercialization in just one year, and so has the independent networking industrial chain. Throughout its development history, the popularization of China's 5G applications has been global leading.

South Korea started to apply 5G technology relatively early. By March 2021, the cumulative number of 5G users in the country reached 14.48 million, and the penetration rate of total mobile users was 20.4%. On the one hand, the 5G traffic effect is outstanding. At present, the traffic of 5G users in South Korea exceeds that of 4G users. The average monthly traffic of 5G users is around 25–26 GB, and the distribution of traffic usage is relatively even.

On the other hand, South Korean operators have launched a large number of application services, actively exploring the application innovation of 5G in the fields of industrial Internet, medical care, smart transportation, culture, robotics, and urban public safety and emergency. Taking VR/AR as an example, by the end of 2020, the number of types of application experiences of LG U+, a South Korean telecom operator, reached 4,800.

Meanwhile, the US has been advantageous in staying technologically ahead. In March 2020, the White House released the National Strategy to Secure 5G of the United States, proposing to accelerate the domestic deployment of 5G, assess the risks associated with 5G infrastructure and determine its core security principles, and promote responsible 5G global development and deployment.

As per data from Omdia, the number of 5G users in the US in 2020 was small. It was only 9.9 million, lower than the previous forecast due to the impact of COVID-19. Although the penetration of 5G among users in the US is not as fast as that of China and South Korea, the US stays relatively advanced in the fields of millimeter wave and dynamic spectrum sharing (DSS). The millimeter wave has been commercialized in many cities such as New York, Los Angeles, and Chicago. Meanwhile, the three major American operators have already made millimeter-wave 5G commercial services available.

In general, despite 5G network infrastructure construction in various countries having suffered certain impediments due to the pandemic, its progress has been relatively stable, and the development of 5G exhibits a positive state.

5G Changes Production

It is believed that 4G has changed people's lives and enabled human society to enter the all-IP era. And 5G is more powerful than 4G – it has changed production, which is why 5G technology is highly valued by all countries.

5G technology has the characteristics and advantages of the Internet of Everything, high speed, ubiquitous network, low latency, low power consumption, and safe reconstruction. Its development has profoundly changed production. 5G builds the core basic capabilities of the Internet of Everything, which not only brings faster and better network communication, but shoulders the historical mission of empowering all walks of life.

In terms of sensory multi-dimensional interaction functions, 5G's broadband feature can support more perceptual intelligence such as hearing, vision, and touch. It will promote the full implementation of smart technologies such as AR/VR, holographic video, and tactile Internet. Taking VR/AR/MR as an example, VR, AR and MR (mixed reality) provide users with an immersive video experience through "re-contextualized" information, thus completely changing the transformative technologies for traditional man-machine interaction.

On the demand side, VR/AR/MR technologies emphasize visual, tactile, auditory, and other multi-sensory interactive modes, which agrees with the development tendency of natural consumer behavior needs; on the supply side, high-quality enterprises strengthen their layout and optimize user experience, while lower product prices and richer content attract attention from the user community again.

5G, with the natural advantages of mobility and access anytime and anywhere, can provide more flexible access for VR services, enabling them to move from fixed scenarios and access to mobile scenarios and wireless access. 5G+Cloud+VR/AR/MR can perform real-time processing of complex rendering programs by placing them on the cloud server through the 5G network. It reduces the requirements for the GPU and other hardware, promotes multiple application scenarios, and attains more convenient and more vivid communication

Moreover, 5G will free devices from wired connection for good, thus truly making them wireless and lightweight, and optimizing user experience. It is foreseeable that in the future, immersive experience + smart space will become the most obvious feature.

In terms of the ubiquitous deployment of computing power, the unique architecture of 5G can realize distributed computing, meaning there is a longer limitation to physical concentration or embedded hardware. The seamless connection between the cloud and the terminal will enable the infrastructure to evolve into informatization and intelligentization. In the era of the Internet of Everything, every device can directly access the cloud server and communicate with it efficiently; a sea of information will enter the cloud server network and continue to "feed" AI. This means far greater application efficiency of cloud servers and much faster learning progress of AI. The China Academy of Information and Communications Technology predicts that by 2025, the size of China's cloud computing market will exceed 500 billion yuan, and more than 80% of enterprises will migrate key tasks to the cloud.

In terms of real-time data circulation, a powerful 5G network can build integrated scenarios covering satellites, the IoT, etc. to promote larger and wider data collection, transmission, storage, processing, and application, to accelerate data flow and transform it into a value stream. The IDC predicts that by 2025, the data scale of China will reach 48.6ZB, with a compound growth rate of more than 30%.

The IoT is widely used in the healthcare industry. Taking telemedicine as an example, through the IoT, doctors can perform remote diagnosis and treatment of patients regardless of the distance. However, limited by the transmission rate under the 4G network, during telemedicine, there are some common problems like low-definition images, which make it difficult for the doctors to identify a patient's condition. The development of telemedicine services is slow, and it has suffered great restrictions during the promotion. The integrated application of 5G and the IoT has broken through some of the limitations of 4G, improving the image quality and alleviating the delay of information transmission.

More Evolution

To realize massive real-time information exchange and an immersive metaverse experience, requires continuous improvements in computing power. However, at present, 5G construction still needs to greatly improve its communication capabilities. Regardless of the active promotion of 5G application development in every country, the current application is still faced with immature global industrial standards, imperfect industry convergence technologies and standards, cross-field application development gaps, etc. There is still a long way to go for industrial development.

Though major powers in the world are actively promoting 5G network upgrades based on the R15 version of the SA (stand-alone) architecture, the SA terminal is still not mature enough, and network slicing and edge computing technical solutions still require further improvement. Although the R16 standard has been frozen, it will take some time for mobile communication technology to leap from standard formulation to equipment R&D, network upgrade, terminal popularization, and wide application. This is the basic law of technological and industrial development, and the development of 5G technology and industry is going to be a long-term process too.

Simultaneously, technical standards related to industry applications still need raising. On the one hand, the development and application of 5G capabilities need to be closely coordinated with technologies such as the IoT, cloud computing, and AI. The introduction of software-defined, virtualized, cloud-based, and open 5G new technologies may bring new security risks. On the other hand, the vertical industry has its own shortcomings: the technology and ecological maturity that support the development of 5G fusion applications such as high-definition video and AR/VR need to be improved, issues such as 8K coding and decoding, intelligent driving algorithms, and industrial scene application models wait to be resolved, and the standards for typical application scenarios have to be formulated quickly.

In terms of network, 5G coverage needs to be widened. Moreover, the rapid development of 5G fusion applications requires the wider deployment of such networks. At present, the coverage effect of SA base stations is limited, and 5G network based on NSA fails to support some massive connections and low-latency scenarios. With the growing demand for 5G from industry, existing networks are clearly unable to meet their needs. It is urgent to continue to explore new 5G construction models such as network slicing, private networks, and intelligent and simplified networks for further development.

In terms of industry, there are still weak links in the 5G industrial chain. The core communication links such as radio frequency chips, medium and high frequency devices, and the industrial foundation are weak. They need both government and industry assistance.

There are also development bottlenecks in both the personal and industrial terminal markets. Under the promotion of new information consumption and the economic development of an inner loop in China, the market lacks typical applications such as Douyin and WeChat in the 4G era. Despite the many types of personal consumption terminals, none of them are super popular. Moreover, users do not fully understand 5G, the industry's digital capability is insufficient, and services that adapt to it are still to be developed. On the basis of a deep understanding of 5G, broadband services, and industry predicament, it is necessary to jointly explore solutions while breaking the old profit distribution model of traditional industry to create new business models. 5G has given us a peek of the metaverse, and the 6G under active R&D is going to lead us to truly walk into it.

2.3 Artificial Intelligence: Brain of the Metaverse

At present, intelligent tools collect, transmit, process, and execute information data and other subjects of labor. While the instruments of labor in the past have effectively extended human limbs, the combination today in the information society has broken the limitations of the human brain. It is a revolution that strengthens and extends human intelligence and liberates humans from rigorous work.

Today, AI has become an important driver for the new scientific and technological revolution, and industrial transformation. And the breadth and depth of its role are comparable to those of previous industrial revolutions. AI is the focal point of the current scientific and technological revolution: it connects knowledge and technical capabilities in various fields intelligently, and releases the huge energy accumulated. In the metaverse, it will also play the quintessential role of an intelligent and innovative brain.

Metaverse Administrator

Before AI became the administrator of the virtual world, it has gained recognition in administering the real world. A smart city is a comprehensive carrier for the final manifestation of AI application scenarios.

Intelligence is commonly believed to be a characteristic of creatures (humans) with vital signs and many physical body perceptions. Therefore, a smart city is like a city that has been made alive. In fact, the city itself is the result of continuous life growth, while the "smart city" is an evolving concept.

A smart city is used to refer to a digital city. As the idea took root and evolved in a broader urban scope, people began to understand that a smart city is essentially a better quality of life and more efficient use of resources through the intelligent application of information. Its ultimate goal is sustainable urban development.

The growth of cities is always closely related to technological expansion. From the city once imagined to the city seen with the eyes, and to the "walking city" proposed by British architect Ron Herron, cities are shifting from static to dynamic, as the Internet, IoT, cloud computing, and big data have made everything more convenient. The urban state where modern technologies are integrated is part of the smart city concept

The technical core of a smart city is smart computing, which has the potential to connect various industries, such as urban management, education, medical care, transportation, and public utilities. A city is the carrier of all industries. Therefore, smart computing is the technological source of smart cities. It will affect all aspects of urban operations, including municipal administration, construction, transportation, energy, environment, and services.

Despite the different definitions of a smart city in academia, the six dimensions of a smart city proposed by professor Rudolf Giffinger of Vienna University of Technology in 2007 are generally acknowledged, namely: smart economy, smart governance, smart environment, smart people, smart mobility, and smart living.

Smart economy includes innovative and entrepreneurial spirit, economic image and trademark, industrial efficiency, labor market flexibility, international network embeddedness, and technical transformation ability.

Smart governance includes decision-making participation, public and social services, transparency of governance, political strategies, and perspectives.

Smart environment includes pollution control, environmental protection, and sustainable resource management.

Smart people include educational background, lifelong learning, social and ethnic diversity, flexibility, creativity, openness, and participation in public life.

Smart mobility includes local accessibility, a (international) barrier-free communication environment, good communication technology infrastructure, sustainability, innovation and safety, and transportation systems.

Smart living (quality of life) includes cultural facilities, health status, personal safety, residence quality, educational facilities, tourist attractions, and social harmony.

These six dimensions comprehensively cover all aspects of urban development. In addition to the material elements of a city, social and human elements are also included, and high quality of life and environmental sustainability are set as important goals. In other words, to make cities smarter, the key lies in the effective use of information and communication technology. The ways to achieve this are improving the economy, enhancing the environment, and strengthening urban governance. Related to urban space is also the improvement of transportation (mobility) efficiency.

As with AI being the brain of cities, when it rises to the metaverse, it assumes the role of metaverse administrator as well. Obviously, based on real-time super-large feedback, to ensure operation and content supply efficiency, it is necessary to assist the administration of the metaverse system through multi-skilled AI. It is difficult to maintain a complex system like the metaverse by relying solely on manpower. Therefore, similar to the role of NPCs in video games, AI will support the daily operation of the metaverse in the future.

Multi-skilled AI collects and processes information in a more human-like way by combining functions such as computer vision, audio recognition, and natural language processing. It is AI that can adapt to new situations and solve more complex problems. Therefore, in the future, AI will assume both the front-end service responsibilities of the metaverse such as customer service and NPC, and the back-end operational responsibilities, such as information security review, daily data maintenance, and content production. Moreover, as computing power and technologies advance, the operation and content supply efficiency of the metaverse will be guaranteed.

Innovative Content

At present, with stronger underlying computing power and increasingly abundant data resources, the empowerment of AI for various application scenarios keeps transforming various industries. For such a huge system as the metaverse, the content richness will be far beyond imagination. Moreover, the content will be provided to users in the form of real-time generation, experience, and feedback. The requirements for supply efficiency will greatly exceed human capabilities, and more mature AI is needed to empower content production, make dreams come true, and lower the threshold for content creation.

The border of the metaverse keeps expanding to meet ever-growing content demands. Also, AI-assisted content production/full AI content production is required. Only with content production empowered by AI can such content demand be met.

In fact, whether it is a traditional online game or a blockchain game, game scripts have always been the most important factor in destroying the game economy. Game players harvest game resources by playing the games, while game scripts generate game resources through automated execution, multi-tasking, and so on, which shrinks the labor value of game resources. The automated game script exploits the player's labor, and the development AI is expected to completely replace the player's mechanical labor in the games, and even their intellectual activities such as PVP.

GPT-3, which achieved a breakthrough in 2021 as a mainframe computer model for learning human language, has 175 billion parameters. It applied deep learning algorithms to train itself through thousands of books and a host of texts on the Internet, and eventually managed to imitate human writing. However, current AI models have not yet been able to directly understand semantics and texts. Therefore, in the short term, AI will mostly assist with content production. By simplifying the content production process, it enables creators to turn their imagination into reality, thus lowering the threshold of content creation. But with the further advancement of AI and machine learning, full AI content production is expected to materialize in the future, thereby directly satisfying the ever-growing demands of the metaverse.

2.4 Digital Twins: Embryonic Form of the Metaverse

Technically, we have seen the embryonic form of the metaverse – the digital twin.

At SIGGRAPH 2021, the leading academic conference on computer graphics, Nvidia, a well-known semiconductor company, released a clip of a 14-second speech given by digital doubles during the press conference of Nvidia CEO, Jen-Hsun Huang, in April 2021.

Though it was only 14 seconds, his iconic leather jacket, expressions, movements, and hair were all synthesized, which fooled almost everyone. It shocked the industry. As one of the foundations of the metaverse, it is clear that the digital twin is developing rapidly. Metaphorically, as a dynamic simulation of the real world, the digital twin is a tentacle of the metaverse extended from the future.

Conceptual Evolution

In the digital age, the digital twin, as one of the most important digital technologies, is irreplaceable in the digitization of human society, and therefore is frequently discussed in keynote speeches at major summits and forums, attracting great attention both inside and outside the industry. With the maturity of the idea and the advancement of technology, a world of digital twins is under construction – from components to complete machines, from products to production lines, from production to services, and from static to dynamic.

The digital twin concept was born in the United States. In 2002, Michael Greaves, a professor at the University of Michigan, proposed the concept of virtual digital expression equivalent to physical products: a digital replica of a specific device or group of devices that can abstractly express the real one/ones and be used as a basis for testing under real or simulated conditions. The concept stems from the desire to more explicitly express device information and data, hoping to gather all the information for more advanced analysis.

This concept was put into practice in NASA's Apollo program, which took place before the proposal. In this program, NASA had to build two identical spacecrafts – the one that stayed on Earth was a twin, which was used to reflect (or mirror) the state of the spacecraft that was performing the mission.

Nowadays, many mainstream companies in the industry have given their own interpretations and definitions of a digital twin, but in fact, our understanding of it keeps evolving. This is seen in Gartner's discussion of digital twins over the past three years: in 2017, he defined a digital twin as a dynamic software model of an object or system, through which billions of physical objects will be expressed within three to five years. In the emerging technology maturity curve released by Gartner in 2017, the digital twin was in the budding stage of innovation, and 5–10 years away from mature applications.

In 2018, he defined it as a digital expression of a real-world object or system. With the wide application of IoT, digital twins can connect real-world objects, provide information on their status, respond to changes, improve operations, and increase value. In 2019, his definition changed to a digital mirror image of an object, process or system in real life. Large systems such as power plants or cities can also create their own digital twins.

Horizontally, from the perspective of model requirements and functions, some people believe that a digital twin is a three-dimensional model, a copy of physical entities, or virtual prototypes. On the data dimension, some believe that data is the core driving force of digital twins, focusing on its value in product life cycle data management, data analysis and mining, data integration and fusion, etc.

On the connection dimension, some believe that a digital twin is an IoT platform or industrial internet platform; this view is focused on perceptual access from the physical world to the virtual world, reliable transmission, and intelligent services. For services, there is one opinion that the digital twin is a simulation, virtual verification, or visualization.

As mentioned above, there are many different interpretations of the digital twin – with no unified definition. However, it is generally accepted that physical entities, virtual models, data, connections, and services are the core elements.

Broadly speaking, a digital twin creates a digital virtual model based on the physical entity of a device or system. This virtual model will render services on the information platform. It is worth mentioning that, unlike the computer-assisted drawings, the biggest feature of the digital twin is its dynamic simulation of the physical object. In other words, it "moves."

Meanwhile, the basis for the "movement" of the digital twin is the physical design model of the physical object, the "data" fed back by the sensors, and the data of historical operation. The real-time state of the object and the external environmental conditions will be "connected" to the "twin."

From Virtual Reality Mapping to Full Life Cycle Management

The digital twin, as a concept beyond reality, is based on the core elements of a digital mapping system, and the matching social needs, of one or more important and interdependent equipment systems. The popularity of the digital twin is rising in the digital age; this is another tentacle of the metaverse extended from the future.

The real-virtual mapping is the fundamental feature of the digital twin. It realizes bidirectional mapping between the physical model and the digital twin model by constructing a digital twin model for the physical entity. This plays an important role in improving the performance and operation of the corresponding physical entities. For future intelligent fields such as industrial Internet, smart manufacturing, smart cities, and smart medical care, virtual simulation is a necessary link. The basic characteristics of real-virtual mapping pave the way for industrial manufacturing, urban management, medical innovation, etc., to transform from "heavy" to "light."

Taking the Industrial Internet as an example, in the real world, to repair an enormous piece of equipment, it is necessary to consider matters like profit and loss of shutdown, the complex equipment structure, and to arrange personnel to conduct on-site inspections. It is obviously a "heavy project." With the digital twin, repairmen only need to feedback data to the digital twin to judge the condition of the actual physical equipment and complete the inspection and repair.

With the concept of the digital twin, GE proposed one way to integrate physical machinery and analysis technology, and applied it to the manufacturing of its aero engines, turbines, and MRI equipment. Thus, every piece of equipment has a digital twin, realizing accurate monitoring, fault diagnosis, performance prediction, and control optimization.

During the COVID-19 pandemic, the world-famous Leishenshan Hospital was built using a digital twin. The Central-South Architectural Design Institute (CSADI) was ordered to design the second Xiaotangshan Hospital in Wuhan – Leishenshan

Hospital. The building information modeling (BIM) team of CSADI created a digital twin for it. According to the project requirements, BIM was also used to guide and verify the design.

The construction of digital twin cities in recent years has ignited transformative innovations in intelligent urban management and services. For example, the Xiongan New Area in Hebei, China, has an amazing digital twin of an urban underground integrated pipe system. It integrates 12 types of municipal pipelines, including underground water supply pipes, renewable water pipes, hot water pipes, and electric power communication cables; the digital twin city of Yingtan in Jiangxi Province won the Global Smart City Digital Transformation Award at the Barcelona Global Smart City Conference.

Additionally, since the real-virtual mapping is a dynamic simulation of physical objects, it means that digital twin model continues to grow and enrich itself: throughout the product life cycle, from product demand, functional, material, service environmental, structure, assembly, process, test, to maintenance information, it continues to expand, enrich and improve.

The more complete the digital twin model is, the more it can approximate its corresponding entity, so as to visualize, analyze and optimize it. When the various digital twin models of the product life cycle are compared to scattered pearls, the chain connecting these pearls is the digital thread. The digital thread can connect not only the digital twin models of various stages, but also information on the full product life cycle, thus ensuring the consistency of various product information when changes occur.

In the field of the full life cycle, Siemens has extended the value of digital twins to many industries with the help of the tool, Product Lifecycle Management (PLM), and has achieved remarkable results in the fields of medicine and automobile manufacturing.

Taking GlaxoSmithKline's vaccine R&D and production laboratory as an example, through the comprehensive construction of digital twins, the complex vaccine R&D

and production have finally achieved a total virtual whole-process twin monitoring. The company's quality control expenditure decreased by 13%, there was 25% less rework and scrap, and it saved 70% on compliance costs.

From real-virtual mapping to full lifecycle management, digital twins cover a wide range of application scenarios for various industries. As an Internet giant that is leading the march into the metaverse, Microsoft has adopted digital twins in the bottom layer of its detailed technical layering of commercial metaverse applications. Obviously, the digital twin based on multiple digital technologies is the most specific understanding of the metaverse.

The mixture of virtual reality and social form constructed by the metaverse is more like a fusion of the digital twin and real physical space in a strict sense, and we can shuttle between the real world and virtual worlds at will.

CHAPTER 3

Building the Metaverse Economic System

Tim Sweeney, founder of Epic Games and father of the Unreal Engine, once commented that the metaverse would be more pervasive and powerful than anything else – so that when a central corporate controls it, the corporate will become more powerful than all governments and rule the Earth. The metaverse is a gigantic platform. To prevent the monopoly of such a centralized platform, it is imperative to establish the economic rules of the metaverse. And blockchain is an important technology that offers a solution for value delivery.

Having experienced the evolution from a single decentralized ledger application to a value delivery layer in virtual space and time, blockchain has set a value delivery model in the virtual world. With the open-source application ecology and innovative business model, blockchain applications are developing rapidly and prospering, setting off fast iterations on a global scale.

From Bitcoin to Ethereum, and to the recently popular DeFi and NFT, blockchain has proved itself to be an efficient clearing and settlement platform across time and space. Its emergence ensures the circulation of virtual assets can be decentralized and independent, and that the rules are fair and transparent through open-source code, while the emergence of smart contracts and DeFi maps real-world financial behavior to the digital world.

3.1 Block and Chain

Technological Integration of Blockchain

Blockchain is a great invention in the history of human science, yet the blockchain seen by the public today is not an entirely new technology. In fact, it includes research findings in multiple fields in different historical periods. In 1969, the Internet was born in the United States, and thereafter the Internet has expanded from four American research institutions to the entire globe. In the nearly five decades since – from the earliest application in military and scientific research to all aspects of human life – there have been five technological breakthroughs that were particularly significant to blockchain development.

(1) TCP/IP PROTOCOL

In 1974, the TCP/IP protocol – the core communication technology of the Internet jointly developed by American scientists Vinton Cerf and Robert Kahn – was released officially, thus determining the position of blockchain in the Internet's technical ecosystem. This protocol made it possible to transmit information between different computers and different network types. It set a unified information dissemination mechanism for the Internet. All computers connected to the network can communicate and interact as long as they follow it.

The TCP/IP protocol is of great importance to both the Internet and blockchain. Since its invention, the entire Internet has been relatively stable among the underlying hardware devices, and the intermediate network protocols and network addresses. However, at the top layer, innovative applications continued to emerge, including news, e-commerce, social networking, QQ, and WeChat, as well as blockchain.

Put another way, in the technical ecosystem of the Internet, blockchain is a new technology of the top layer – the application layer. Its emergence, operation, and development have not affected the underlying infrastructure and communication protocols of the Internet – it is still one of the many software technologies that operate according to the TCP/IP protocol.

(2) CISCO ROUTER TECHNOLOGY

The router technology invented by Cisco in 1984 is a mock object of blockchain. Leonard Posak, director of Stanford University's computer center, and Sandy Lerner, director of the computer center of Stanford University's business school, co-designed a networking device called the "multi-protocol router" (MPR). They put it into the communication lines of the Internet to help the data travel from one end of the Internet to the other (thousands of kilometers away) accurately and quickly.

In the entire Internet hardware layer, there are tens of millions of routers directing the transmission of Internet information. An important feature of Cisco routers is that each router saves a completed Internet device address table. Once it changes, it will be synchronized to tens of millions of other routers (theoretically) to ensure that each router can calculate the shortest and fastest path. For routers, even if a node device is damaged or hacked, the transmission of the entire Internet information will not be affected.

(3) B/S (C/S) ARCHITECTURE

The B/S (C/S) architecture comes from the World Wide Web, which is abbreviated to WWW, and divided into WWW clients and WWW servers. All updated information is only modified on the WWW server, while tens of thousands, or even tens of millions of other client computers, do not retain the information. The information data is obtained only when the server is accessed. This structure is the B/S architecture of the Internet – the central architecture. This architecture is also the most important architecture of the current Internet. Internet giants, including Google, Facebook, Tencent, Alibaba, and Amazon have adopted it.

The B/S architecture is of great significance to the blockchain. It only allows data to be stored in the central server, from which all other computers obtain information. However, in blockchain, tens of millions of computers have no center (decentralization), and all data is synchronized to all computers. This is the core of blockchain.

(4) Peer-to-peer network (P2P)

The peer-to-peer network is another Internet infrastructure corresponding to C/S (B/S). It is characterized by multiple computers connected to each other in a P2P position. A computer can be both a server – setting shared resources for other computers in the network – and a workstation.

Napster, as one of the earliest P2P systems, is mainly used for music resource sharing; yet it cannot be regarded as a real P2P network system. On March 14, 2000, a message was posted on the Slashdot mailing list – an American underground hacker site – claiming that AOL's Nullsoft division had released Gnutella, an open-source clone of Napster. In Gnutella's distributed P2P network model, each networked computer is functionally equivalent – both a client and a server – so Gnutella was called the first true P2P network architecture.

Blockchain is a software application of P2P network architecture. It is a benchmark application of P2P networks that have broken out from the silence of the past.

(5) Hash algorithm

The hash algorithm is an algorithm that converts a number of random lengths into a fixed-length value via a hash function. It is essential to the functioning of the entire world. From Internet application stores to email, to antivirus software, browsers, and so on, everything adopts the secure hashing algorithm. It can tell if Internet users have downloaded what they want and if they're the victims of a man-in-the-middle or phishing attack.

The generation of new coins from Bitcoin or other virtual coins adopts the function of the hash algorithm to obtain numbers that meet the format requirements, and then the blockchain program gives Bitcoin rewards.

Mining, including Bitcoin and tokens, is a math game built with a hashing algorithm. However, because of competition ferocity, people all over the world use powerful servers for calculation to get the rewards first. A great number of computers on the Internet are in this math game.

Blockchain is essentially a decentralized database able to perform distributed recording and storage of data information. It is also a data structure that combines the blocks into a chain! It adopts cryptography to generate a set of time-sequential, immutable, and trustworthy databases. This uses decentralized storage, can ensure data security, and builds a consensus among participants on the chronological order of records and the current state of transactions made on the entire network.

Since 2017, the concept of blockchain has been a hit. Much of the media has tried to help people understand what blockchain is all about in plain words. In layman's terms, blockchain changes the past one-person bookkeeping to a mode where everyone can perform bookkeeping; thus accounts and transactions are more secure. This is distributed data storage. In fact, in addition to distributed storage, technical terms related to blockchain also include decentralization, smart contracts, encryption algorithms, etc.

The blocks in the blockchain are data blocks generated using cryptographic methods. Data is permanently stored in the form of electronic records. The files that store such electronic records are called blocks. Each block records several items, including the Fibonacci Number, block size, block header information, number of transactions, and transaction details. Each block consists of a block header and a block body. The block header is used to link to the address of the previous block and ensure integrity for the blockchain database; the block body contains verified transaction details or other data records during the block creation.

Data storage of blockchain ensures the integrity and rigor of the database in two ways: first, the transactions recorded on each block are all value exchanges that take place after the previous block comes into being and before the block is created, which safeguards the integrity of the database; second, in most cases, once a new block is completed and added to the end of the blockchain, the data record of this block can no longer be modified or deleted, guaranteeing database rigor.

The chain structure relies mainly on block header information between each block to link, and the header information records of the previous block (the hash value converted by the hash function), which is recorded in the next new block. This completes the information chain of all blocks.

In the meantime, due to the time stamps on the blocks, blockchain is also time-sequential. The older a blockchain is, the more blocks are linked, and the more expensive it is to modify the blocks. Blocks employ a cryptographic protocol that allows a network of computers (nodes) to co-maintain a shared distributed ledger of information without requiring full trust between nodes.

This mechanism guarantees that the information stored in the blockchain can be trusted as long as the majority of the network is published to the blocks according to the stated administrative rules. This ensures that transaction data is replicated consistently across the network. The presence of a distributed storage mechanism usually means that all network nodes save all information stored on the blockchain. Metaphorically, the blockchain is like the crust of the earth – the deeper, the older, the more stable, and the less change.

As the blockchain records all block transactions since the genesis block was invented, and the data records cannot be tampered with, all value exchange activities between both transaction parties can be tracked and found. Such a completely transparent data management system is not only impeccable from a legal point of view, but provides a credible tracking shortcut for existing logistics tracking, operation log records, and auditing.

When a new block is added to the blockchain, there is a tiny probability of a "fork," that is, two blocks that meet the requirements appear simultaneously. The solution to a "fork" is to extend the time, wait for the next block to be generated, and select the longest branch chain to add to the main chain. The probability of one "fork" is tiny, and the probability of multiple forks is basically zero. And "fork" is only temporary. The final blockchain must be the only determined longest chain.

From a regulatory and auditing perspective, entries can be added to but not removed from the distributed ledger. A network of communication nodes that runs specialized software replicates the ledger among participants in a P2P fashion, performing maintenance and verification of the distributed ledger. All information shared on the blockchain carries an auditable trail, which means it carries a traceable digital "fingerprint." Information on the ledger is pervasive and persistent, and by creating a reliable "transaction cloud," data will never be lost. Thus, blockchain fundamentally eliminates the risks of single-point failure and data fragmentation discrepancies between counterparties.

In general, the blockchain holds six technical characteristics: decentralization, openness, autonomy, anonymity, programmability, and traceability. It is these six technical characteristics that make it a revolutionary technology.

Decentralization: as distributed accounting and storage technologies are adopted, there is no centralized hardware or management organization, and the rights and obligations of any node are equal. The data blocks in the system are jointly maintained by nodes with maintenance functions in the entire system, and the stoppage of any node does not affect overall system operation.

Openness: the system is open – except the private information of the transaction parties is encrypted – and blockchain data is so open that everyone can search for it and develop related applications through the open interface.

Autonomy: the blockchain adopts consensus-based specifications and protocols, thus enabling all nodes to exchange data freely and securely in a credible environment. This shifts the trust in "people" to trust in machines and technology.

Anonymity: as exchanges between nodes follow a fixed algorithm, their data interactions do not require endorsement from trust, and counterparties do not need to disclose their identities.

Programmability: the digital nature of a distributed ledger means that blockchain transactions can be linked to computational logic and are inherently programmable. Therefore, users can set algorithms and rules that automatically trigger transactions between nodes.

Traceability: the blockchain stores all historical data since the genesis block through the block data structure, and any piece of data on the blockchain can be traced back to its origin through the chain structure.

Once the blockchain information is agreed and added to the blockchain, it is jointly recorded by all nodes, and is guaranteed to be correlated through cryptography. It is highly difficult and costly to tamper with it. In fact, the blockchain seems to be productive, but as a decentralized self-organization, it has more characteristics of the emerging production relations.

From Bitcoin to Smart Contracts

From blockchain 1.0 to blockchain 3.0, it has witnessed a great forward leap. The first blockchain simply referred to Bitcoin's general ledger, which records all transactions that have occurred since the Bitcoin network began operation in 2009. From an application point of view, the blockchain is a secure global general ledger via which all digitized transactions are recorded.

On October 31, 2008, Satoshi Nakamoto (pseudonym), the founder of Bitcoin, published a paper – Bitcoin: A Peer-to-Peer Electronic Cash System, in which he claimed to have invented a new electronic currency system not controlled by governments or institutions. He has clarified the Bitcoin model, pointing out that decentralization, non-additional issuance, and infinite division are its basic features. And blockchain is the basis for its operation.

In January 2009, Satoshi Nakamoto released an application case of blockchain – the open-source Bitcoin software on the SourceForge. Through "mining," he obtained 50 Bitcoins, naming the blockchain that generated the first Bitcoins the "genesis block." One week later, Nakamoto sent 10 Bitcoins to cryptographer Hal Finney, completing the first Bitcoin transaction in history. Thereafter, the Bitcoin frenzy swept the globe.

The first Bitcoin exchange was set up on February 6, 2010. And on May 22, someone bought two pizzas with 10,000 Bitcoins. (By 2021, 1 Bitcoin was worth over $30,000!) On July 17, 2010, the famous Bitcoin exchange Mt.Gox was established, marking its true entry into the market. Nonetheless, it was mainly geeks crazy about Internet technology who could understand Bitcoin and engage in this crypto trading: they discussed Bitcoin technology on forums, mined Bitcoins on their own computers, and traded them on Mt.Gox.

Bitcoin has decentralized preparation, production, recording, and circulation. In the Bitcoin network, multiple parties maintain the same blockchain ledger, and determine the bookkeeping rights through "mining," – calculating random numbers, so as to make the ledger decentralized, secure, and non-tamper-able. Stimulated by the economic rewards in "mining," "miners" will voluntarily purchase "mining equipment" to provide the necessary computing power to maintain the entire transaction network and ensure system security.

Having gone through more than ten years of verification, Bitcoin has managed to convince some overseas market institutions and governments to accept its value storage function. At present, its circulation market value has reached about 930 billion dollars, and the system's computing power is about 18 OEH/s (1.8×10^{20} hash calculations per second). After years of operation, there has never been a serious security issue, and more and more people have accepted its asset properties. Though it cannot circulate as legal tender, it acts as a general equivalent among some currency holders. Such transfers are also not based on any centralized account system.

The success of Bitcoin proves that decentralized value transfer can come true. Based on such success, Ethereum borrowed and upgraded its model to support more complex program logic. Eventually, it created smart contracts, thus taking the blockchain from version 1.0 of a decentralized ledger to 2.0 – a decentralized computing platform.

At the end of 2013, Vitalik Buterin founded Ethereum, the earliest digital token ecosystem. Ethereum is a blockchain-based smart contract platform. It is the "Android system" on the blockchain: everyone can use Ethereum services to develop applications on the system. Now, thousands of application "buildings" have been built on the transformed Ethereum system.

Ethereum was designed to create a blockchain 2.0 ecosystem. Essentially, this is a public blockchain platform with complete Turing scripts known as the "world computer." In addition to value delivery, developers can also create smart contracts on Ethereum, which has expanded blockchain commercial channels, such as the issuance of tokens for many blockchain projects and the development of decentralized DAPPs (distributed applications).

Ethereum has realized decentralized general computing through smart contracts and virtual machines. Its developers can freely create decentralized applications and deploy contracts. While controlling the mining, Ethereum miners need to execute the contract program through the virtual machine, and generate new blocks from the new data states. Other nodes must verify whether the contract is executed correctly while checking the blockchain, so as to ensure the credibility of the calculation results.

A smart contract in Ethereum is a preset instruction that always behaves as expected. The concept of smart contracts was first proposed by Nick Szabo in 1995. They allow trusted transactions without a third party that are traceable and irreversible. The smart contracts on Ethereum are open and transparent and can be transferred, realizing trust through the open source. However, if hackers discover and exploit a program loophole first, there will also be a loss of assets.

In short, by carrying smart contracts, Ethereum writes a certain agreement between A and B into the program in the form of "If-else," and invites the entire network to witness it. The smart contracts will be automatically executed when they expire, thereby avoiding the need for additional friction costs from a centralized witness. At present, the ecological popularity of Ethereum continues to expand. Recently, an average of 250 new contracts have been deployed daily, with an average of 160,000 contracts daily as well. And the numbers continue to grow.

The DAPPs based on blockchain's smart contracts mainly focus on finance, games, gambling, and social networking. The number of users and amount of assets is growing steadily. DAPPs have achieved the decentralized execution of key logic through smart contracts on the chain, solving trust problems in some scenarios, such as credit transfer in financial applications and key values in game applications. Different from traditional network applications, DAPPs do not require registration but use a decentralized address to identify user information.

DeFi (Decentralized Finance) is the most active DAPP. It replaces financial contracts with smart contracts and provides a series of decentralized financial applications. Users can complete financial operations related to virtual assets through DeFi, and use it to reconfigure virtual assets in terms of capital allocation, risk, and time.

DeFi reproduces a financial system on the blockchain by programming financial contracts. Applications on DeFi can be roughly divided into stablecoins, debit and credit, exchanges, derivatives, fund management, lottery, payment, and insurance. In reality, many DeFi functions far exceed these nine categories, mainly because they can be combined with each other like Lego blocks. Because of this, they are also known as Lego money.

DeFi is efficient, transparent, threshold free, and freely composable. These characteristics have allowed the DeFi ecosystem to expand rapidly and prosper. More users have accepted it and are using it. And anyone can access and use DeFi. According to a research report by Fabian Schär, published on the official website of the Federal Reserve of St. Louis, DeFi can improve the efficiency, transparency, and accessibility of financial infrastructure. In addition, the composability of the system allows anyone to combine multiple applications and protocols to create new and exciting services.

DeFi has far-reaching significance for the metaverse: users have total control over all asset financial activities on their own chains, and owners' financial operations are free from the classical restrictions of geography, economy, and trust. Through smart contracts, black box operations can be avoided. The combination of DeFi and NFT can expand to the content, intellectual property rights, records and identification, financial documents, and so on of the metaverse. This creates a transparent and autonomous financial system that can accommodate more diverse assets and more complex transactions. Ultimately, such combinations support the civilization building of the metaverse.

Put another way, blockchain is an inevitable outcome of Internet big data technology progression. The ever-growing number of users and ever-expanding data are bound to create higher requirements for information security. It is foreseeable that in the metaverse era, the current blockchain will also be further upgraded to meet even higher demands.

3.2 NFT

Blockchain is merely an underlying technology – a new application mode of computer technology, including distributed data storage, point-to-point transmission, consensus mechanisms, and encryption algorithms. It is like everyone's mobile phone, while Bitcoin is just one of its APPs. At present, blockchain is becoming a high-performance commercial application, having built the underlying architecture and overcoming technical difficulties at all levels. When this is fully realized, we will enter the era of blockchain 3.0.

There is no doubt that the metaverse will become the biggest application in the blockchain 3.0 era. And the emergence of NFT has assetized virtual items, becoming the forerunner of blockchain entering the metaverse.

Digital Item Ownership

NFT is short for non-fungible token – indivisible and unique. The birth of the NFT is based on an Ethereum pixel icon project from 2017, called CryptoPunks. The total amount of these pixel icons was capped at 10,000, and every pixel icon must be different. Those who owned an Ethereum wallet could collect the pixel icons of CryptoPunks for free at the time, and could trade their own pixel icons in the secondary market.

It should be noted that most assets in the real world and the virtual world are non-fungible. Therefore, NFT allows items such as art, collectibles, and even real estate to be tokenized. There can only be one formal owner at a time, and such owners are protected by the blockchain, where nobody can modify ownership records or copy/paste new NFTs. In other words, it can determine the ownership of digital items in the virtual world at a low cost, thereby laying the foundations for more economic activity in the metaverse.

Primarily, NFTs can map virtual items and become the transaction entity of virtual items, thus turning them into assets. All data content can be mapped on the chain through links so that the NFT becomes its asset "entity" and completes its value flow. By mapping digital assets, equipment, decoration, and land property rights can all become tradable entities.

In other words, NFTs can bring metaverse rights to fruition, enabling humans to create a true parallel universe in the blockchain world. Like physical keys, programs can confirm users' authority by identifying their NFT. This helps create the decentralized transfer of rights in the virtual world, meaning the transaction of virtual property rights can occur without the presence of a third-party registration agency. NFTs propose solutions by essentially providing a data-based "key" that can be easily transferred and exercised. Moreover, a series of corresponding permissions can exist outside the centralized service or centralized database. This greatly enhances the efficiency of data asset transactions and circulation, while its circulation requires no third-party participation at all.

In the field of collectibles, the digital scarcity that comes with NFTs is ideal for collectibles or assets whose value depends on limited supply. And Crypto Kitties and Crypto Punks are two of the earliest NFT use cases. One single Crypto Punk NFT – Covid Alien – sold for $11.75 million! In 2021, popular brands like NBA TopShot were trying to create NFT-based collectibles that contain video highlights from NBA matches rather than still images.

In the field of art, an NFT enables artists to sell their work in natural form without having to print it out. Also, unlike physical art, artists can make money through secondary sales or auctions, ensuring their original work is recognized in subsequent transactions. Markets are dedicated to art-based NFTs. For example, Nifty Gateway 7 sold/auctioned over $100 million of digital art in March 2021.

In the gaming field, NFTs also open big windows of opportunity in ownership. While people spend billions of dollars on digital game assets, such as avatars or outfits in Fortnite, consumers don't necessarily own these assets. NFTs will allow players of crypto-based games to own assets, earn them in games, port them outside the games, and sell them elsewhere (like open markets).

In the virtual world of CryptoVoxels, holding the NFT of a certain plot means holding the right to develop, transform, arrange, and lease the limited plot space. The system does not record the user's permission information in the server, but the corresponding NFT's. The NFT in CryptoVoxels can be regarded as an advanced form of the land deed: its circulation and execution do not require an intermediate registration agency; the ownership and transformation rights are transferred through the on-chain token, and users who own the NFT can directly obtain the corresponding permissions.

Secondly, the emergence of NFTs will also alter the business model of virtual creation. Virtual goods will change from services to transaction entities. In the traditional mode, game equipment and game skins are a kind of service by nature rather than assets. They are theoretically unlimited, and the cost to make them approximates to zero. Operators usually sell game items to users as service content rather than assets, and the same happens to creation platforms – where users have to pay specified fees when using other works. The existence of NFT has altered the conventional trading mode of virtual commodities. Users/creators can directly produce virtual commodities and trade them, just like in the real world. NFTs can be separated from the game platform, and users can freely trade related NFT assets.

The assetization of NFT equities in the metaverse can promote the circulation and trading of such equities. This distinction allows any rights in the metaverse to be easily financialized, such as access, viewing, approval, and construction rights. It facilitates the circulation, rental, and trading of such rights.

2021: First Year of NFT

2021 was the year of the NFT. Since 2021, the market volume of NFT art, sports, and games have been increasing. Speculators and cryptocurrency enthusiasts flocked to purchase this new type of asset, which stands for ownership of online-only items such as digital art, trading cards, and online worlds. According to Coin Gecko's data, in 2021 H1, the total market value of NFTs was up to $12.7 billion – an increase of nearly 310 times compared to 2018. As per data from Non Fungible, the 2021 Q2 NFT trading volume scale reached $754 million – a year-on-year increase of 3,453% and a month-on-month increase of 39%. The growth was explosive.

For example, on March 11, 2021, *Everydays: The First 5000 Days* was sold for nearly 70 million dollars ($69,346,250). The creator of *Everydays: The First 5000 Days* is Beeple, an American digital artist and graphic designer. Beeple has drawn one digital picture every day since May 1, 2007 for thirteen and a half years non-stop. Having compiled them together, he published *Everydays*.

The successful bidder will receive all digital drawings of *Everydays:* and an NFT, which, based on blockchain, stores the metadata of the digital work, the painter's signature, and history of ownership. Also, it's a one-off – a true collector's item. Christie's will send an NFT that represents the ownership of *Everydays: The First 5000 Days* to the successful bidder's cryptocurrency account.

For another example, on August 27, 2021, NBA star Stephen Curry updated his Twitter icon to a BAYC NFT in a tweed suit, for which he paid $180,000 (55 ETH, worth around RMB 1.16 million). It attracted great attention on the market. The full name of BAYC is Bored Ape Yacht Club, which is a collection of 10,000 ape NFTs, including 170 different rare items, such as hats, eyes, expressions, clothing, and backgrounds. Through programming, 10,000 one-of-a-kind apes were generated randomly, each with a different expression and outfit.

The popularity of NFT has also drawn the capital market to the game. Among them, OpenSea took advantage of its own NFT users and NFT asset types to quickly dominate the market share of NFT exchanges. In August 2021, its NFT transaction volume exceeded $1 billion, accounting for 98.3% of the global total. Comparatively, its full-year transaction volume was lower than $20 million in 2020.

One reason for the explosion of NFT transaction volumes is the fast enrichment of supply-side content; the number of NFT projects has surged. In August 2021, the number of NFT projects with a total transaction volume exceeding 1 ETH reached 2,776 – over three times that of January 2021 (less than 700). Also, the number of Axie Infinity uses represented by game scenes, and of CryptoPunks and Bored Ape Yacht Clubs represented by social scenes, skyrocketed. On August 28, 2021, Axie Infinity tweeted that the number of daily active users on its Android version reached 1.01 million (passing 1 million for the first time). Its Windows version surpassed 380,000, and its Mac version was approximately 23,000, while its iOS version was about 15,000.

On the other hand, OpenSea's dominance comes from its easy entry to NFT platforms and fees lower than most competitors. There are no restrictions on creators entering OpenSea. Creators can simply apply for an account on OpenSea and publish their own NFT products. It sets a low threshold, whereas its competitors all require an application or an invitation to participate in the release of NFT products or transactions.

The handling fee of OpenSea is 2.5%. Though this is significantly higher than that of the conventional cryptocurrency market, it is the lowest in the industry. For other NFT trading platforms, it is usually 10% or 15%. Also, OpenSea charges lower royalties from creators. In addition, in order to ensure the uniqueness of creators, OpenSea must not distribute the royalties to multiple addresses; NFT creators cannot use part of their royalties directly for other purposes through OpenSea.

NFTs are the product of blockchain and metaverse development. In the future, more capital and giant companies will enter the NFT market. An almighty scramble for market share and control is about to break out.

3.3 Solving Key Metaverse Problems

Blockchain is an important technology that connects the metaverse concept. Based on its own technical characteristics, blockchain naturally adapts to the key application scenarios of the metaverse. Blockchain is a traceable chain data structure that combines information blocks, which continue to be generated in a sequential manner. It is a distributed ledger that cryptographically ensures that data cannot be tampered with or forged. With its own characteristics, blockchain can be applied to digital assets, content platforms, game platforms, the sharing economy, and social platforms. In a way, it is the bridge that connects the bottom layer and the upper layer of the metaverse.

Virtual Assets and Virtual Identities

The problems with users' virtual assets and virtual identities on traditional Internet platforms hinder the arrival and development of the metaverse. For example, the right to interpret traditional Internet virtual property is often held by platforms, and the nature of such property is not clear; the economic system of the virtual world is completely dependent on the operation level of the operators, which makes it difficult for unprompted adjustment and balance; user identities and derived data are completely in the hands of the platforms, which means low privacy levels.

The blockchain can ensure the user's virtual asset rights and interests are not controlled by a single organization through the decentralized record of such rights and interests. This type of recording makes virtual assets similar to real assets in the physical world; users can freely dispose, circulate, and trade them without the restrictions imposed by centralized institutions.

The mature DeFi blockchain ecosystem can supply an efficient financial system for the metaverse. From the mortgage, securitization, insurance, and other aspects of virtual assets, it renders low-cost, low-threshold, and high-efficiency financial services to users. Their virtual assets, which are made of nothing different from real assets, are entitled to financial services, further strengthening the asset nature of virtual items. Through the stability of virtual property rights and a rich financial ecosystem, the economic system of the metaverse will have the adjustment function of the real world, and the virtual value created through the user's labor will be determined by the market.

Also, blockchain makes it easier for traditional virtual assets to circulate across multiple platforms. Virtual assets, including traditional game assets, are recorded in operating agencies' databases. Their cross-platform transfer requires the mutual data trust of multi-parties, which is costly and difficult to achieve. The ownership information of virtual assets is recorded through NFTs, which is then traded in a P2P manner in the blockchain's decentralized network. In nature, these projects use the blockchain platform to clear and settle assets as well as lowering risks while improving clearing efficiency.

The blockchain technical breakthrough which enables users to control their own identity data has been found at last: W3C proposed the concept of DID (distributed digital identity) based on blockchain. DID is characterized by security, autonomy, controllability, and portability. Based on DID, the role of social network applications is one of a service provider which cannot monopolize social data. Also, social network links between people will not happen at the data level but at the application level. This mode can effectively promote new social applications to adapt to the complex and diverse social scenarios of the metaverse.

Curb the Impropriety of Centralized Platforms

Centralized platforms can hurt users through asymmetric advantages over rules. In the Internet age, centralized platforms often use their own traffic advantages and asymmetric advantages to exploit platform users to a certain extent. With the concealment of rules in Internet application services, they make gradual adjustments to serve their own ends; thus, overall profit is tilted towards these platforms – like big data discrimination.

From the age of the Internet to the age of metaverse, this enormous platform for human virtual activity naturally monopolizes Internet traffic. The metaverse business model dominated by centralized platforms will inevitably lead to a larger monopoly and control, which is more problematic than Internet monopoly, and not conducive to the long-term development of the metaverse. Therefore, to prevent such a monopoly, a balance of three important factors – decentralization, security, and efficiency – must be struck.

Blockchain is the key to solving this: its structure is essentially a chain of events connected chronologically, and all transactions since the genesis block are recorded in the blocks. Accounting information such as transaction records will be packaged into blocks and encrypted, and timestamped simultaneously; all blocks will be connected into a general ledger in their timestamp order.

The blockchain adopts the cryptographic mechanism specified in the protocol for authentication to ensure that no tampering and forgery can occur. To modify the ledger record in the blockchain, one has to crack and revise the encrypted data on the entire chain, which is extremely difficult. And this is determined by the structure of the blockchain.

Another factor ensuring blockchain security is distributed storage. Even if the hacker cracks and modifies a node's information, it is in vain. Only by simultaneously modifying data of more than half of the system network nodes can the data be truly tampered with. The cost of such tampering is extremely high – almost impossible – thereby ensuring blockchain security.

The blockchain builds a set of protocol mechanisms. It enables every network node to verify the results recorded by other nodes. Only when most nodes (or all nodes) in the entire network consider the record to be correct simultaneously, or when all participating nodes compare results and pass them on unanimously, can the authenticity of the record be recognized by the entire network. Then, the recorded data be written into the block.

Blockchain adopts distributed data storage to solve account disaster recovery, and at the same time, establishes a P2P relationship between individuals to form a decentralized data system. Without a central organization, all nodes have the same rights and obligations, and the stoppage of any node will not affect the overall operation. Therefore, one of the key advantages of distributed storage is decentralization.

Blockchain naturally makes it possible to achieve a balance: user assets and user information do not have to be recorded on the platform that provides the content, but can be encrypted and recorded on the underlying blockchain platform. In this mode, the content platform cannot monopolize user information, and does not have the right to interpret user virtual equity either.

In addition, smart contracts have the advantages of permanent operation, data transparency, and immutability. First, the number of nodes supporting the blockchain network is often hundreds or even thousands; the failure of some nodes will not halt the smart contract. Secondly, all data on the blockchain is open and transparent, so the data processing of smart contracts is the same, and any party can view its code and data at runtime. Finally, all blockchain data cannot be tampered with; thus the smart contract code and the data output cannot be tampered with. Nodes that run smart contracts never have to worry about malicious code and data modification by other nodes.

The biggest plus points of smart contracts are the automation of relevant procedures, reduction of personnel involvement, and improved efficiency. Through smart contracts, the blockchain will truly decentralize the operation of platform rules.

3.4 Unfinished Blockchain

Blockchain is tremendously valuable in building the metaverse economic system. According to the wooden barrel theory, the capacity of a barrel is determined by its shortest stave. Therefore, the factors that restrict the development of blockchain lie in some problems that have surfaced in its development. They affect the implementation of the blockchain and also challenge its current development.

Firstly, as the blockchain advances, the volume of blockchain data stored by nodes will grow. Therefore, the storage and calculation burden increases, making it difficult for core clients to operate. While lightweight nodes can partially solve this problem, industrial-grade solutions suitable for larger scales remain to be discovered.

Secondly, the application efficiency of blockchain is low. A Bitcoin transaction requires six confirmations, each of which takes about ten minutes, meaning the entire network confirmation takes about an hour. Such efficiency is not suitable for high-performance (millisecond) financial transactions, like stock trading. As the blockchain advances, the efficiency problem can be addressed in certain ways: for example, alliance chains and private chains can greatly improve transaction performance by reducing nodes and optimizing algorithms. Meanwhile, under consensus mechanisms such as DPoS or PBFT, the transaction confirmation is fast, and the transaction volume can also satisfy the expected transaction scale and most business needs.

Thirdly, blockchain decentralization is not completely reliable. Blockchain works well on the premise of excellent credit from both parties in the default transaction. However, in actual transactions, this premise cannot be fully guaranteed. Moreover, where there is a credit problem with one party, the transaction cannot be revoked in time; this loophole will lead to serious social and economic order issues.

In traditional services, there is a central organization that acts as the business intermediary body in such situations. When a security threat occurs, the business body only needs to release the corresponding security patches to deal with the issue. Blockchain, however, is built on the basis of consensus. In theory, there is no way to solve security problems unless all participants are on board. In Blockchain 2.0, the advent of self-agreed smart contracts exacerbates this risk. Some consensus directly covers up the hidden dangers of subsequent security, and there are some smart contracts without modification mechanisms. Therefore, it is necessary to establish strict accountability mechanisms and regulatory systems to ensure the safety and legality of transactions. It will take considerable time to establish and implement a complete set of mechanisms.

Fourthly, there should be a balance in the privacy protection of the blockchain. In the blockchain's public chain, every participant can obtain a complete data backup, and all transaction data are open and transparent. This is both an advantage and a disadvantage of blockchain. The solution to the privacy protection of Bitcoin is to achieve anonymity by isolating the relationship between the transaction address and the real identities of the address holders. However, the transaction itself is so public that everyone can access transaction information in Bitcoin or the system, which is strictly prohibited in industries such as medicine and finance.

In the alliance chain, in addition to processing algorithms, there are other special ways to protect privacy data. The Enigma system, for example, breaks data into pieces and runs some clever math to mask them. It should be noted that privacy protection will affect certain transaction performances; therefore, the two need balancing.

The ultimate goal of the metaverse is to build an ecosystem that is independent yet connected to the real world. Therefore, a sound and transparent currency system will be a prerequisite to ensure operation. Though the trust system built by the blockchain will become the infrastructure of the metaverse and the foundation of its economic system, for now, there is still a long way ahead. As aforementioned, in the real times of the metaverse, the blockchain applied must have evolved from today's version.

CHAPTER 4

Virtual Technology Paves the Critical Path

As the information revolution marches forward, the human pursuit of "creating another world" has continued to make great progress. Today, the development of graphics, multimedia, human-computer interaction technology, and brain science has paved the way for the birth of the virtual world. Virtual technologies represented by VR, AR, and MR are driving the human world into the metaverse.

4.1 Evolving XR

XR (extended reality), which consists of VR/AR/MR, not only covers the spectrum between complete reality and complete virtuality, but allows these technologies to be collectively referred to as a content range. This is an important and promising component in the development of the metaverse.

Virtual Technology Integration

Virtual technologies use computer software and hardware, and various sensors (such as high-performance computers, graphic generation systems, special clothing, gloves, glasses, etc.) to generate a lifelike three-dimensional simulation environment. Through a variety of special-purpose devices, they enable users to immerse themselves in the environment and directly interact with it in a natural and simple manner.

Virtual technologies enable users to inspect or operate objects in the virtual world while providing them with visual, auditory, and other natural real-time perceptions. Whether it is VR, AR, or MR, as a branch of virtual technology, it is inseparable from the support of three major technology clusters: stereoscopic displays, 3D modeling, and natural interactions.

(1) STEREOSCOPIC DISPLAY TECHNOLOGY

Stereoscopic display technology is based on the principle of stereoscopic vision of the human eyes. Therefore, it is necessary to study and master this mechanism, so as to design the stereoscopic display system. Next, certain technologies can be applied to achieve the stereoscopic effect through the display device.

Stereoscopic display technology can be subdivided into HMD (head mounted display) technology, holographic projection technology and lightfield imaging technology. The basic principle of HMD technology is that after the image is reflected through a prism, it enters the human eye and is displayed on the retina. This creates the effect of viewing an ultra-large screen at an ultra-short distance with high enough resolution.

Holographic projection technology can be divided into projection and reflection, which is the reverse display of holographic photography. Different from the principle of traditional stereoscopic display technology that uses a binocular parallax, holographic projection technology displays 3D images by projecting light on air or special media (such as glass and holographic film). People can view the image from any angle and get the exact same visual effect as in the real world. At present, the holographic projection technology applied in various performances requires the use of special media, and various precise optical arrangements need to be made on the stage in advance. The effects of this type of performance are dazzling, but the cost is high, the operation complicated, and the operators must be well-trained.

In a sense, lightfield imaging technology can be seen as quasi-holographic projection technology; it forms an image with a spirally vibrating optical fiber, with the light bouncing directly from it to the human retina. Simply put, it uses optical fibers to directly project the entire digital lightfield to the retina – producing the so-called cinematic reality.

(2) 3D MODELING TECHNOLOGY

3D modeling is achieved through 3D software, 3D scanning, and lightfield capture. 3D software modeling constructs a model with 3D data in a virtual 3D space through various 3D design software. This model, also known as 3D model, can be presented as a 2D flat image through 3D rendering, computer simulation, or 3D printing equipment.

When building a virtual reality world, in addition to the adoption of conventional 3D modeling technology and real-life shooting technology, 3D scanning can also be applied to quickly model the real environment, characters, and objects, and convert their stereoscopic information into digital models for processing. A 3D scanner is a tool that adopts 3D scanning to quickly build digital models of real-world objects or environments. The many types of 3D scanners can generally be divided into two categories: contact and non-contact.

Lightfield capture modeling was first applied to Lytro – a company founded by Ren Ng in 2006. It captures images from multiple optic angles by placing a micro lens array in front of a single sensor; however, this plan greatly compromises the resolution. In recent years, companies such as Facebook Reality Labs, Microsoft MR Studio, DGene, Shenzhen Prometheus Vision Technology, and WIMI Hologram, have employed a multi-camera array of hundreds of cameras and one depth camera to create a ring capture system, which shoots the object in all directions. Through high-speed processing, AI algorithms, and a dynamic fusion system, the 3D model is synthesized in real time.

(3) NATURAL INTERACTION TECHNOLOGY

To achieve perfect immersion in virtual reality, the support of natural interaction technology is critical. Interaction technologies such as motion capture, eye tracking, voice interaction, and tactile interaction are tremendously helpful.

Obviously, in order to achieve a natural interaction with the scenes and characters in virtual reality, the basic movements of the human body need to be captured, including gestures, facial expressions, and body movements. The mainstream technologies for gesture recognition, expression, and motion capture are divided into two categories – optical motion capture and non-optical motion capture. Optical motion capture includes active and passive modes, while non-optical motion capture includes inertial, mechanical, electromagnetic, and ultrasonic motion capture.

Eye tracking uses cameras to capture human eye or face images, and detects, locates, and tracks human eyes or faces via an algorithm, so as to estimate the change of the user's line of vision. At present, two image processing methods – spectral imaging and infrared spectral imaging – are commonly used. The former captures the contour between the iris and sclera, while the latter tracks the contour of the pupil.

During real world interaction, in addition to eye contact, expressions and actions, there is also voice interaction. A complete voice interaction system includes two major parts: speech recognition and semantics comprehension. Usually, when speech recognition is mentioned, it refers to both. Speech recognition involves three technologies: feature extraction, pattern matching, and model training. It covers the fields of signal processing, pattern recognition, acoustics, auditory psychology, AI, etc.

There have always been related applications of haptic interaction technology, a.k.a. force feedback technology, in the game industry and virtual training. Specifically, it gives users a more real immersion by applying some force, vibration, etc. It enables the creation and control of virtual objects in the virtual world, such as remote control of machinery or robots, and the simulation training of surgical residents to perform operations.

In short, virtual technology directly takes us to a virtual 3D space, blending us in with the interactive environment. In the virtual world, we can move freely and enjoy the scenery, just like in the real world. There, we have sufficient autonomy.

VR/AR/MR

At present, VR/AR/MR is an essential part of virtual technologies. XR, which consists of VR/AR/MR, not only covers the spectrum between complete reality and complete virtuality, but allows these technologies to be collectively referred to as a content range.

XR is divided into multiple layers – from a virtual world with limited sensor input to one with full immersion, or from a mixed world built through the superposition of auxiliary equipment to one that can be fully perceived by the naked eye. It enables physical objects of the world and digital objects of the virtual world to coexist and interact, delivering an effective fusion at last.

VR: a collection of simulation techniques, and computer graphics man-machine interface technology, multimedia, sensing, network technologies, etc. It is also a cutting-edge subject and research area of selected interdisciplinary technologies. Its three major characteristics are immersion, interaction, and imagination.

AR: a new technology that can "seamlessly" integrate real world information and virtual world information. Through computers and other technologies, it simulates, superimposes, and applies the real information (vision, hearing, taste, touch, etc.) that is difficult to experience within a certain scope of time and space in the real world to the real world. Therefore, it's perceived by humans, and a sensory experience beyond reality is obtained. AR includes multimedia, 3D modeling, real-time video display and control, multi-sensor fusion, real-time tracking and registration, and scene fusion. It has three major characteristics, namely: information integration of the real and virtual worlds, real-time interactivity, and adding and positioning virtual objects in a 3D space.

MR: MR is a further development of VR. By presenting virtual scene information in real scenes, it builds an interactive feedback loop in the real world, the virtual world, and for users to enhance lifelikeness. MR includes AR and VR, and refers to new visual environments that result from integrating the virtual and the real. Also, it can place virtual objects in the real world. Users can see the real world, as well as virtual objects, and interact with them.

4.2 Marketing VR

The Internet and social platforms can neither accurately project the virtual world to the physical world, nor give humans a sense of deep experience in the virtual world. However, VR technically solves both problems. As a computer simulation technique that enables people to enter and experience the virtual world in an immersive way, it is fully capable of creating a vivid space.

Having evolved, VR has now entered the stable production stage. At the same time, VR-based applications and equipment have begun to appear in education, media, entertainment, medical care, heritage protection, and many other fields. After a brief rocky period, VR development and prospectus are again strong; today, VR products are more cautious but more successful than ever.

Another Market Explosion

The concept of VR was in fact around decades ago. It budded as early as the 1960s: the earliest VR was traced back to Sensorama in 1956, which integrated a 3D display, a scent generator, a stereo speaker, and a vibrating seat. There were six built-in short films for people to watch. However, its huge and impractical size prevented it from becoming a commercial, recreational facility.

In 1989, Jaron Lanier formally proposed the concept of VR. In the 1980s and 90s, NASA successively launched rudimentary VR equipment such as experimental HMD, earphones, and gloves. In 1985, NASA developed an LCD optical head-mounted display, which could provide an immersive experience under the premise of miniaturization and light weight. From then on, its design and structure have been widely promoted and adopted. However, due to the limitation of chip and processing technology at the time, expensive professional equipment was needed. Therefore, it was not popularized for the civilian market. Instead, it was only used in military training, aircraft manufacturing, aerospace, etc.

In the games and entertainment field, some well-known companies have also tried to use VR to develop related products. In 1993, gaming company Sega planned to develop a head-mounted VR device for consoles; however, it failed due to mediocre performances in internal testing. In 1995, Nintendo released the Virtual Boy – a console based on VR. Yet it flopped on the market, partly due to its limitation to only red and black colors as well as low resolution and refresh rate.

What truly revived commercial VR was the release of the Oculus Rift and Google Glass in 2012. Since then, VR products have significantly improved in cost, latency, field of view, and comfort. And commercial VR equipment has also entered the market, pushing the VR industry back into the boom times.

2016 was another milestone year for VR equipment, and for the content ecosystem. As VR is included in a number of national policies in China, such as the informatization of the 13th Five-Year Plan, Chinese manufacturers have entered the game. LeVR, GoVR Player, InLife HandNet VR, Deepon VR, etc., have appeared one after another. At CES 2016, Oculus officially released the Oculus Rift HMD, along with HTC Vive and Samsung's GearVR. Since then, financial capital has been less skeptical over the VR content (film, game, etc.) market; a large number of investments are pouring in.

And emerging Chinese game companies and VR studios have launched some high-quality VR works, such as Eternity Warriors VR and Aeon.

Despite 2016 being dubbed "the first year of VR," a "cold winter controversy" broke out at the end of the year. According to a Canalys report, in the first quarter of 2017, American consumers made up 40% of global VR market sales; Japan rose to second, reaching 14%; China's market share dropped to 11%, receding to third place.

The quintessential reason was that when most companies – targeted at hardware such as glasses and head mounted displays – rushed to build platforms and portals, insufficient content hindered the interlocking chain of the VR industry. Meanwhile, VR pursues an immersive and scene-based experience, but because user participation is still too limited, the number of users cannot maintain public enthusiasm for VR.

In 2018, driven by the market and technology, the VR industry welcomed a period of recovery and growth. Meanwhile, with the support of 5G, the cooperation of all parties in the industry chain and telecom operators has also accelerated VR development. Consequently, VR continues to be applied in films, education, training, shopping, commodity experience, medical care, transportation, security, ecological protection, and other industries.

The COVID-19 pandemic in 2020 has further sped up the penetration of VR. During the pandemic, the world's first 5G+VR "Cloud Cherry Blossom Appreciation" – jointly created by Xinhua News Agency, Wuhan University, and China Mobile – made cherry blossoms in Wuhan University frequent the trending list on social media. In addition, with the finalization of Mount Everest's height measurement, China Mobile took the lead in opening a 5G+VR slow live-streaming of Mount Everest at an altitude of 6,500 meters, and built a "cloud stage" on that mountain. With VR live streaming, the "snow mountain party" came true.

Oculus Takes the Lead

The VR market boom continues: according to Industrial Securities, the global VR HMD shipments were 6.7 million units in 2020 – a year-on-year increase of 72%; the number is projected to reach 18 million in 2022. In 2020, the number of global VR users surpassed 10 million, and by 2025, it will reach 90 million. And the successive entry of giants such as Apple has brought more possibilities to the market.

(1) Oculus

In terms of market share, Oculus takes the lead: its Quest 2 topped the chart for consecutive months. According to data released by Steam, in March 2020, the top four brands of SteamVR were Oculus, HTC, Valve, and Microsoft WMR. Oculus ranked first with a remarkable share of 58.07%. And the market share of the Oculus Quest 2 has soared since its release. In February 2021, it was the No. 1 VR HMD on Steam. In March, its share continued to grow to 24.25%, topping the SteamVR chart for two consecutive months as the most active VR device.

As Facebook's latest-generation VR all-in-one machine, the Oculus Quest 2 has achieved impressive market performance. Its pre-orders were 5 times that of its last generation when it was released in September 2020. According to Andrew Bosworth, vice president of Facebook Reality Labs, within half a year of its release, its cumulative sales have already exceeded the sum of all previous Oculus VR HMDs.

According to SuperData statistics, the Oculus Quest 2 sold 1.000 million units in a single quarter of 2020 – Q4; according to Yivian's conservative estimate, it sold about 2.5 million units in 2020, and since 2021, the sales volume has been close to 1.5 million units; the cumulative number has been approaching 4 million.

On the investment day on November 17, 2021, the CEO of Qualcomm said that the sales volume of Oculus Quest 2 under Facebook had reached 10 million units.; Mark Zuckerberg disclosed that the VR platform had set 10 million users as a remarkable milestone. And crossing it means the arrival of the sustainable development. Judging from the strong sales of Quest 2, the Facebook VR ecosystem has taken shape, and users will continue to contribute to content revenue in the future.

According to SuperData, Quest 2 shipments accounted for 87% of all standalone VR devices in 2021. And thanks to strong Quest 2 sales, Facebook's non advertising revenue surged in 2020 Q4, achieving $885 million (including its hardware products Oculus and Portal) – rapid year-on-year growth of 156%. The hot sales of Quest 2 are the main driving force for such growth.

(2) HTC

HTC and Valve have co-developed VR HMDs with a rich product line, such as HTC VIVE. The first developer version of VIVE was released at MWC in 2015, while its consumer version was made available in 2016. According to Steamspy, the product sold nearly 100,000 units in its first three months.

VIVE Focus Plus supports 2K resolution, 6 degrees of freedom control, and Inside-Out positioning tracking. It can be used without connecting to a PC or a tracker, thus greatly improving user experience. VIVE Pro has higher color contrast, built-in 3D stereo earphones, and a spatial location tracking function within 100 square meters, which can meet the needs of large games. The latest VIVE Cosmos has the highest resolution in the VIVE series and is compatible for a variety of VR applications.

According to IDC statistics, HTC's share in the global VR industry sales revenue in 2018 Q1 reached 35.7%; Samsung was next with a share of 18.9%. In terms of content, as of March 2021, there were 3,871 exclusive VR games on Steam. Of these 3,727 support HTC VIVE and 2,708 support Oculus Rif, which is an outstanding advantage.

4.3 VR Life

As the hardware interface between the metaverse and the real world, VR has become an important application for games, videos, and live stream. It is empowering downstream industries faster. And it has been widely used in real estate transactions, retail, home decoration, cultural tourism, security, education, medical care, etc. IDC forecasts that in the future, as the VR industry chain continues to improve and rich data accumulates, VR will be fully integrated with various industries, thus casting a strong flywheel effect.

Meanwhile, as VR matures towards mass production, it still faces many development barriers; there is an arduous and long road ahead for it to fully realize its potential and achieve market dominance.

VR in Daily Life

VR video: commercialization has taken place, and immersion, interactivity and content innovation are expected to improve. Early VR videos were mainly short landscape clips. With the maturity of shooting techniques, VR IMAX theater, VR live streaming, and VR 360° video have gradually been realized; they have broken through venue and screen size limitations, and extended the original scenes to the AR/VR end, providing users with diversified content such as movies, TV series, and sports events. In 2020, iQiyi launched the 360° panoramic 8K VR interactive theater *Killing a Superstar*, which enables users to explore different scenes in the same time dimension through interactions. It was a brand-new VR narrative mode.

VR game: *Half-Life: Alyx* is a VR-exclusive RPG game developed by Valve (unlisted), which integrates multiple game elements, such as VR, FPS, and puzzle solving. Its high image quality and strong physical interaction create a deep sense of immersion. As *Half-Life: Alyx* comes free with the purchase of VR equipment Valve Index, SuperData reveals that as soon as the game was announced in 2019 Q4, Valve Index (which charges $999) was sold out in 31 countries. Its annual sales in 2019 reached 149,000 units, while sales in 2019 Q4 made up about 70%. On March 24, 2020, the game was released on Steam. It received 10,654 positive reviews on the first day! The applause rate was over 95%.

VR shopping: VR comprehensively integrates online and offline scenarios, restores shopping mall scenarios through modeling, empowers the traditional economy via digitalization, improves shopping convenience, and leads the new remote consumption trend. In 2020, China Telecom launched the 5G SA+MEC XR digital twin platform to provide consumers with business scenarios such as virtual shopping guides, virtual landscapes, and red packet hunting. It also supports business owners with big data analysis, creating an open digital ecosystem and thereby laying the foundation for VR shopping.

VR education: compared with traditional classrooms, smart classrooms and labs based on VR enable students to interact with virtual objects and abstract concepts, deepen their knowledge memory, and create a high-quality and immersive teaching environment. For example, in lab scenarios, some experiments are too dangerous, and teaching equipment is expensive. VR can easily solve such problems.

VR tourism: according to the Tourism Cities Federation (TCF) and the Tourism Research Center of the Chinese Academy of Social Sciences, in 2019, there were 1.5 billion international tourists, with a global tourism revenue of $5.8 trillion. The market space is vast. And VR-based travel schemes, compared with traditional offline travel, can save users time and cost. Users can enjoy the beautiful sceneries of different places at different time periods around the world without leaving home. At present, Huawei Cyberverse is able to perform the independent explanation and restoration of cultural relics. In the future, as technology advances, VR + tourism is expected to commercialize even faster.

VR medical care: in terms of medical training, VR can break through conventional production and funding limitations, lower medical education costs, create an efficient learning environment, and help medical students to repeatedly train skills on virtual operating tables; in terms of medical intervention, VR can simulate a variety of environments, thus alleviating the anxiety of mentally ill patients; in terms of clinical diagnosis, VR can simulate patient signs, comprehensively obtain patient information, and predict surgery risk points in advance; in terms of telemedicine, VR can cross spatial boundaries, address insufficient medical resources, or go deep into a war zone and severely afflicted areas to give real-time guidance. As hardware technology improves, VR is expected to become an important supplement to traditional medicine.

The Long Road to Consumption

Internationally, VR has matured. It is developing towards a multi-sensory immersive experience that includes vision, hearing, and touch. In the meantime, the corresponding hardware devices are developing towards miniaturization and portability. Obviously, the future promise of VR is great; it is the key path to the metaverse. However, it is undeniable that due to technical, service, and other system imperfections, there is still a gap between VR and the consumer market.

A survey targeted at experts in the US in 2018 shows that among factors that hinder the popularity of AR and VR, user experience was considered the most important, with 39% and 41% of respondents choosing them, respectively. If the device performance is not up to standard, the user experience will be much worse. As the first metaverse entry in the future, AR and VR still require continuous optimization in software and hardware.

In terms of experience, there is still room for improvement in the definition and refresh rate of current VR devices. Contemporary VR product types include VR display docks, VR HMDs and VR all-in-one machines. The best resolution of VR devices on the market is 4K. As mentioned above, to achieve the most natural definition for the human eye, it must be as high as 16K. A high refresh rate can improve the smoothness of the images, reduce latency and double image, and to a certain extent, alleviate the vertigo that people may experience when using VR devices. The ideal refresh rate is 180HZ, while most of the current VR HMDs refresh between 70–120HZ.

Clearly, VR still faces some secondary but critical challenges that directly impact user experience and determine their willingness to use. For example, the heating of electronic components is tricky for wearable devices, because high computing power and high communication bandwidth generate far more heat. Therefore, heat generation and cooling of equipment will become key research areas of subsequent product development. In addition, there is an implied need for customization of glasses-type devices: individual interpupillary distance, short-sightedness, and eye care habits directly affect each consumer's willingness to use the product.

In terms of performance, existing VR devices have a massive computing load and high power consumption, which directly shortens battery life. High performance naturally requires devices to have strong computing power, which can result in excessive power consumption and overheated devices. This is a potential safety hazard. However, it doesn't mean that we have to choose between high performance and low power consumption. One solution is the application of 5G and cloud computing to the VR field. This can effectively release terminal pressure, while reducing their size and weight and improving user comfort.

For example, to achieve a VR display with retina screen effects, the pixel density per unit angle must reach 60 PPD (57.6 PPD to be exact). While this parameter is ensured, it is also necessary to achieve a field of view of 110 degrees and above. Commonly, an optical lens is used to enlarge the field of view. In the narrow space of the VR device, a display component of about 2 inches must have a total amount of horizontal pixels above 6K. In addition, better display, computing, and communication performance require higher energy consumption. Whether part of the performance is placed locally or in the cloud, the improvement of computing power and communication capabilities will consume more energy.

Streaming is another way to reduce computing load. The usual streaming is wired – the HMD is connected to the mobile phone through USB-Type C or to the PC through the DP and HDMI interfaces; thus the main computing power rendered comes from the mobile phone and the PC. Through the cable, the video is directly transmitted, and the interaction information is controlled.

In terms of portability, wireless streaming technology has not yet matured. To make VR HMD lightweight and able to solve spatial movement problems, wireless streaming is necessary. At present, the mainstream wireless streaming technologies are WIFI and private protocols. The former transmits the data rendered and compressed by the PC's GPU to the HMD through the WFI router. Usually, a gigabit router is required for a smoother experience. However, due to technological immaturity, there is additional latency, loss of image quality, high performance consumption, and other unstable factors. The latter performs transmission through the compression algorithm and communication protocol developed by the equipment manufacturer. For example, the VIVE wireless kit, using WiGig accessories, can achieve a delay of less than 7miliseconds between computers and VR HMDs, but it requires an additional WGig accelerator card, which increases user cost.

In terms of price, VR devices are expensive. The price of better-selling VR devices on the market varies, but in terms of comprehensive product parameters, most of the higher end devices charge around 4,000 yuan (around 600 dollars) or more in China. In addition, if the user purchases a VR HMD instead of an all-in-one machine, another host device with satisfactory performance is needed.

Moreover, VR has not yet formed an explicit business model. At present, there are two main ways to profit in the ToC market: one is the sale of terminal equipment and online content payment; the other is the single-payment model of offline VR experience shops. This business model is not only a single structure, but weak in continuity and stability. Hardware and terminal equipment sales are the most important sources of income in the current budding VR market, but content – as the core element of the metaverse – will be the profit cow in the future. At present, the content provided by most VR experience shops is not attractive enough, and customers are basically one-timers, meaning no stable and continuous income to the industry.

The metaverse is driving the overall development of the VR industry into a period of opportunity for technological change – technological innovation is the support, application demonstration is the breakthrough, industry integration is the main thread, platform agglomeration is the center, and the construction of VR+ is the destination, yet the roadmap still requires more work and exploration.

4.4 The Virtual Future – Mixed Reality

While VR is a computer simulation technology that enables people to enter and experience the artificial virtual world in an immersive manner, and one that can create a lifelike virtual world which isolates users from the real world, AR is a further upgrade of VR.

AR can superimpose digital information onto the physical environment. It is a new technology that "seamlessly" integrates real world and virtual world information. Through computers and other technologies, the real information (vision, hearing, taste, touch, etc.) that is difficult to experience within a certain scope of time and space is perceived by humans; thus a sensory experience beyond reality is achieved.

Differences Between AR and VR

VR is a fully immersive technology that presents users with a virtual environment. This makes VR inherently less mobile – a safe environment must be guaranteed for users, where they move within a limited distance to avoid hitting walls or falling.

Since AR superimposes digital objects and information on the real world, its real value to users mainly manifests in mobile scenarios. For example, AR can help users in an unfamiliar environment obtain more information about their surroundings, and navigate them to the destinations. This perfectly integrates AR with the mobile network.

In terms of equipment differences, VR presents a pure virtual scene, and VR equipment often has position trackers, data gloves and helmets, and motion capture systems for users to interact with. In contrast, AR combines virtual and real scenes, and there are 3D cameras on AR equipment. As long as AR software is installed, products with cameras like smartphones can provide an AR experience.

In terms of technical differences, the core of VR is the application of various graphics technologies. Currently, the sense of immersion is most widely used in the field of games and the most pursued, which requires excellent GPU performance. AR emphasizes the restoration of human visual functions, applies many computer vision technologies to 3D modeling and reprocessing real scenes, and values CPU processing power.

In terms of application scenarios, VR characteristics make it immersive and private. This affords it natural advantages in the fields of games, entertainment, education, and social interaction. AR characteristics, however, push it more towards real interaction, suitable for life, work, production, and other fields.

The State of AR: Applications Based on Mobile Phones

In fact, the present consumer-oriented AR HMDs have yet to earn wide market acceptance. However, powered by smartphone OS developer tools such as Android's ARCore, Apple's ARKit, and Huawei's AR Engine, AR has long been popular on smartphones. And AR socializing, games, and navigation have become the most popular applications.

(1) AR SOCIALIZING

Currently, social networking software is undoubtedly the chief application of AR. Snapchat stands out, driving the popularization of AR. During the first quarter of 2021, Snapchat averaged 280 million daily active users, of whom 200 million interact with AR. Its first (by far the most popular) function is to offer users with AR-enhanced filters during video calls. The filters are necessarily practical, but they also make the video calls more fun. For example, users can try out new hair colors and receive feedback from friends.

Brands like L'Oréal take advantage of these filters to advertise their products. Snapchat has also continued to enhance its AR capabilities as it expands. Recognition for other body parts has been added: for example, with foot recognition, users can try on virtual shoes. In addition, users can also apply filters to real-world scenes. These functions give users new experiences and encourage more brands to use AR for advertising and marketing. Many popular video-calling apps are imitating Snapchat's AR function, such as Facebook and Apple's FaceTime. Facebook projected that it would accumulate one billion users of its AR chat filters within three years, including those on Instagram, Messenger, and other product platforms.

Fun AR of Huawei integrates the 3D Cute Moji emoji pack, which can track user facial movements and expressions, and generate his/her 3D virtual icon. Fun AR is widely popular among young users, and ranks among Huawei's top smartphone applications. Apple's FaceTime and TikTok have also adopted similar AR functions. It is mainly through such social networking software that users become familiarized with AR.

(2) AR GAMES

Like AR social apps, games are a form of popular content that introduces AR to the mass market. *Pokémon GO* developed by Niantic is a global success, leading the wave of AR games – it swept the world as soon as it was launched. As of May 2018, *Pokémon GO* had over 147 million monthly active users, and by early 2019 over a billion downloads. As of 2020, its revenue has exceeded 6 billion dollars.

This game is unique because it combines the real and virtual worlds to render players with a real-life AR experience. Pokémon are scattered all over the real world, and players need to move around to capture them. When they discover a Pokémon, it will be displayed in AR as if it existed in the real world. Players can also engage in

Pokémon battles, which are also based on real surroundings (the Pokémon Arena). In addition, the game producer has further combined the game experience with real scenes. For example, players will find aquatic Pokémon near water in the real world.

Pokémon GO is not only a great gaming success, but also an excellent example of novel advertising. That Pokémon is scattered all over the real world and is used to attract people to certain places. In 2016, for example, the game partnered with McDonald's in Japan and the stores were turned into Pokémon Arenas. The partnership attracted an average of 2,000 more customers per day to every McDonald's store. American carrier Sprint also partnered with Niantic to conduct a similar campaign for 10,500 retail stores across the United States. Recently, Niantic's new game, *Harry Potter: Wizards Unite*, teamed up with AT&T, turning 10,000 retail AT&T stores into in-game inns and fortresses to attract customers.

AR games can also be combined with indoor home scenes, such as Nintendo's *Mario Kart Live: Family Circuit*. Players race toy carts equipped with cameras on the track they set up at home, while AR is applied to superimpose graphic elements from traditional Mario Kart games. Only the carts and furniture are real, while other contents are graphic elements superimposed by AR.

Based on HMS Core AREngine, Huawei has co-developed plenty of well-known games with many Chinese Internet entertainment partners (including Tencent, NetEase, Perfect World, Mini Space, etc.). This promotes innovative gaming experiences and the development of the AR ecosystem in China. For example, in *X-Boom*, players are tasked with shooting at AR animal characters displayed in the real world.

(3) AR NAVIGATION

Navigation is another key area of current AR-enabled applications. Both Google Maps and Google Earth have been equipped with AR capabilities. In addition to providing more explicit navigation, "place signs" can be superimposed on real spots such as restaurants or landmarks, so that users can easily access additional information.

Mobile operators are active in the navigation field, too. Able to use 5G positioning, they have a potential advantage over OTT service providers. For example, LG U+ has started the Kakao Navi service, which provides drivers with lane-level navigation that is more accurate than GPS. 5G positioning is also suitable for indoor scenarios, and performs better than GPS-based maps. Moreover, external sensors on AR devices can alert drivers of potential hazards.

Baidu Maps can provide lane-level navigation through difference correction based on the Chinese mobile network. It has been first applied in Guangzhou, Shenzhen, Suzhou, Chongqing, and Hangzhou. Also, tests show that using Huawei HMS Core AR Engine greatly improves the accuracy and stability of Baidu Maps.

Based on AR Engine and HD maps, Huawei Cyberverse has been successfully implemented in Huawei flagship stores, such as Dunhuang, Beijing, and the Shanghai Bund. In addition, this technology has conducted effective explorations in AR navigation, cultural relic reproduction, and better integration of reality and history, thus casting a profound influence.

The capabilities of these AR navigation tools can also give tourists an AR experience. In addition to smartphones, Telefonica has partnered with content producer Mediapro and local transport company TMB to install AR screens on tour buses in Barcelona. And 5G networks can provide location-based live streaming of rich media content as well.

AR + West Lake is an AR tourism innovation application in Hangzhou, China. West Lake is a famous tourist attraction included on the World Heritage List. AR enriches West Lake tourist attractions and provides an immersive viewing experience for tourists. After downloading the app *West Lake on the Palm*, tourists can begin the AR tour. When pointing their mobile phones at the attractions in front of them, the screen instantly displays the background stories related to them. Meanwhile, the app also realizes AR intelligent navigation, provides tour and shopping guides in the entire scenic area, and maximizes tourist convenience.

The Future of AR: MR

Whether it is VR or AR, the ultimate destination of virtual technology is MR. Obviously, MR is the next stage of VR. By presenting virtual scene information in real scenes, MR builds an information loop of interactive feedback from the real world, the virtual world, and users to enhance lifelike experiences.

In short, MR will give people a mixed world – people in the mixed world will be unable to distinguish between digital simulation techniques (display, sound, touch) and reality. Because of this, MR gives more room for imagination. It bit-sizes the physical world in a real-time and thorough manner, while including VR and AR device functions.

In the future metaverse, MR will let players stay connected to both the real and the virtual worlds, and adjust their operations according to their needs and situations. In short, AR and VR are perfectly integrated, and the interaction between them is no longer limited to the imagination.

The final stage of entering the metaverse era will be dominated by MR. Screens will be everywhere in a de-screen sort of way, and there will be a lifestyle that mixes both reality and virtuality, which can be reached anytime, anywhere.

4.5 Holography on the Rise

MR technologies represented by VR, AR, and MR have broken through the perception limit of previous graphic vision, and begun a new stage of 3D experience and interaction. Gradually, vivid 3D virtual dynamic displays and immersive experiences have become the norm. And holography equipped with VR, AR, MR, and other technologies provides a more diversified, realistic, and interactive experience.

As a new technique that depicts, simulates, optimizes, and visualizes the physical world in the virtual world, holography breaks the boundaries among AR, VR, and MR and effectively integrates them. It brings a new, instant, mobile, intelligent, mirrored, and holographic experience mode. From this perspective, the rise of the metaverse is also the rise of holography.

Complete Imaging

Holography means complete imaging.

The idea of holography was first proposed by British scientist Dennis Gabor in 1948. Gabor found that the aberrations created by the lens still store all the object's information. Inspired by the Bragg X-ray microscope, he used coherent electron waves first to record the amplitude and phase of objects. He then used coherent light waves to reproduce an excellent image with corrected aberrations; thus the electron microscope resolution could reach 1A. He deftly transferred the spherical aberration of electronic lenses – which was not easy to correct – to the realm of optics – where such correction is easy – and confirmed his theory with visible light.

With a high-pressure mercury lamp as the light source, Gabor used the direct wave of the transilluminated object as the reference light to interfere with the diffracted wave of the object, obtaining a coaxial hologram. When the hologram was reconstructed via coherent light, the reconstructed image was observed under the microscope. After the coherent beam passes through the hologram, it acquires the phase and amplitude modulation characteristics of the original wave field. The original wavefield appeared to have been released after being captured by photographic disturbance, and the reproduced waveform appeared to have propagated undisturbed. An observer facing the beam found it indistinguishable from the original wave, and the observer seemed to be looking at the original object as if it was still there.

What Gabor saw was an object that holds all the optical properties of the real world; it also holds three-dimensional property and all normal parallax relationships in real life. He solved the fundamental problem in the invention of holography by converting the phase difference into a background wave of intensity difference, thereby encoding the phase into a quantity recognizable by photographic film. Since both the amplitude and phase information of the object wave field was recorded, Gabor called these records a hologram, which means complete imaging.

Now, holography is divided into two major systems: one is in the optical sense, while the other is in the field of projection presentation; but people do not divide the two in practice. Between them, holography (in the optical sense) uses the principle of light interference to record all information in the form of interference fringes by the specific light waves emitted by the entire object, and to form a 3D image very similar to the object itself under certain conditions.

Holography in this sense has three characteristics: first, three-dimensionality – the image reproduced by holography is three-dimensional, as if we are seeing the real object; the second is divisibility – even if the holographic image is broken, it will not affect the image of the entire object, and its integrity will not be lost; thirdly, large information capacity. This suggests that its theoretical upper storage capacity is much greater than the magnetic and optical disks. The holography of projection uses optical transmission to present digital images in the "air," so as to realize the virtual-reality fusion of real objects in the visual space.

Since the concept of holography was proposed, it has been continuously integrated into other disciplines, forming various emerging technologies, such as holographic storage, film pressure micro holography, and holographic metrology. In addition, a variety of holographic techniques have been developed successively, such as transillumination, image plane, rainbow, white light reproduction, true color, dynamic, computational, and digital holography. As more sophisticated and effective holographic techniques are developing, the gradual emergence of holography in society is occurring.

Holography Applications

In recent years, holography has been widely applied in the fashion industry, show business, museums, and political circles.

The combination of holographic film pressure techniques has become a favorite of fashion designers. Back at a fashion week in 2006, designer Alexander McQueen presented a hologram of supermodel Kate Moss. In show business, at the Spring Festival Gala in 2015, Chris Lee performed Shu embroidery (a prominent style of Chinese embroidery from Sichuan Province) with multiple holographic Chris Lees on the same stage. Before that, the sensational Teresa Teng and Jay Chou singing a duet was also achieved through holography.

At present, holographic museums have appeared in some cities in the US and the UK. They exhibit some rare treasures in the form of holograms to reduce the damage and theft of old, cultural relics. There are also designers who have embedded holographic techniques on the back seat of vehicles to provide passengers with information such as menus, communication, and 3D images of the environment. Meanwhile, in political circles, Indian Prime Minister Narendra Modi applied holographic techniques in his election campaign in May 2014 to make himself "show up" in different places and give canvass speeches.

More importantly, holographic techniques equipped with VR, AR, MR, and other technologies will give a more diversified, realistic, and interactive experience. In 2014, Shenzhen Estar Technology Group released the world's first holographic mobile phone, *Takee,* which allows users to view holograms. With a built-in special camera to accurately track the eyeball, it models holographic images. Meanwhile, as the eyeball position continues to change, the picture will automatically adapt accordingly. Users can watch lifelike, high-definition 3D movies with the naked eye without being limited by viewing angles.

In 2015, Microsoft released its holographic glasses *HoloLens.* The glasses have a built-in holographic processor, and internal sensors to sense body movements and convert digital content to project holograms can be used. A floating screen appears in front of the users, allowing them to check the news, make a direct Skype call, watch sport, or play a video game. With *HoloLens*, car buyers can easily choose the color and configuration they prefer, and add and change the functions they desire. Users can also experience common game themes, such as landing on Mars, collecting gold coins, and designing and printing toys in virtual reality. They can even co-design physical models with locomotive designers in Spain. It's all immersive! And Microsoft envisions users sitting in their living rooms and playing holographic 3D games with friends in real-time collaboration. In a way, holography offers a new perspective on seeing the world to human beings.

In education, future students may be able to walk into the learning environment and experience scenes that used to be unreachable just by wearing holographic glasses. This will make learning a traditional, boring text presentation, or a fixed image presentation no longer; instead, contextualized content in words and pictures stimulates greater interest and more effective learning.

New Interactions

In the future, with the help of holographic techniques, a shift from passive to holographic viewing will take place. People can watch, feel, and experience the new environment, and through this "reconstructed" space environment, they can have new interactive experiences. Holography brings a new instant, mobile, intelligent, and mirrored experience mode. People can enter the three-dimensional space created and then interact with the three-dimensional images.

From this perspective, the rise of the metaverse is also the rise of holography. In such an era, the interaction of space and time unifies the five senses, thereby including the viewer's inner feelings and subjective opinions. Participants can walk into this virtual environment created by holographic techniques and interact with it: the sci-fi scenes that used to be imagination only have finally come true.

Picture that in the center of a huge exhibition hall, the hologram of a brand-new concept car floats in mid-air, and details from all angles are explicitly visible. Moreover, as people swipe it with their fingers, the car can be rotated 360°, three-dimensionally disassembled, and even transformed into an airplane! In the future, movies will no longer be played on a screen on the wall; instead, the audience will be on a virtual stage, experiencing or participating in a series of stories.

In fact, this new interactive VR experience is now here. In August 2009, the World Classic Art Multimedia Interactive Exhibition was held in Beijing. All 61 works were amongst some of the finest arts known since the beginning of human civilization: there were paintings and sculptures, including the well-known masterpieces – *Mona Lisa* and *The Last Supper* – by Leonardo da Vinci, as well as the *Venus of Willendorf* dating back to 25,000 B.C. These VR exhibitions not only reproduced all the artworks vividly, but also integrated holographic techniques, 3D technology, and voice interaction into classic art. Thanks to them, these classic works were brought back to life – speaking and moving. The goddess Venus, who is more than 2 meters tall, stood elegantly in the hall, but should you open your arms to embrace her, you will touch nothing! For it was just a hologram in the air.

In the foreseeable future, the unprecedented interaction mode brought about by the metaverse will also open up new spaces for the sustainable development of social civilization and human society. Our understanding of social communication and interaction will undergo a revolutionary change, and the context of information interaction will continue to expand and transform.

CHAPTER 5

From the Games to Beyond the Games

The development of the metaverse is brewing the next technological revolution, but just like every past technological revolution, before one happens, there must be explosive growth in the leading industry. This further drives the development of other elements and promotes the development of related industries. For example, the First Industrial Revolution started in the textile industry, and promoted the development of metallurgy, coal, machinery, and transportation.

At present, the metaverse is seeking an industrial explosion spark to accomplish the spiral rise of. "increase in penetration rate – increase in commercial income – incentivize ecological development – continue to increase the penetration rate." And games, as a more immersive, real-time, and diversified pan-recreational experience, are becoming such a spark, thus accelerating metaverse development.

5.1 The Art of the Game

Since ancient times, games have played a variety of roles in human society. In ancient Greece, games were often a topic of philosophical discussions. Plato, Socrates, Aristotle, and Xenophanes have all explored the meaning of games as part of the system of human thought. From a philosophical angle, these philosophers identified numerous forms of game that can help understand the world and human functioning, and offered three avenues to understand games: competition, imitation, and chaos.

The ancient perception of games was directly related to the gods they believed in. It reflects the relationship between the human and God, or the direct guidance and control of God over man, and the worship of God by man. For example, the winner of a competitive game is considered to have received a favor from the gods; acting or theatre is a series of activities imitating the gods, in order to please them and get closer; speculative games are the choices made by God to guide the direction of the players. Such thinking about games still influences our understanding of them today, pushing gaming development ever deeper and broader.

Play a Game

Bernard Suits creatively proposed the concept of "play a game" in *The Grasshopper: Game, Life and Utopia*, that is, voluntarily overcoming unnecessary obstacles.

Specifically, "voluntarily" is the game attitude, "overcome" requires corresponding game methods, and "unnecessary obstacles" are the rules of the game. In the end, voluntarily overcoming unnecessary obstacles under the game rules achieves the "game goal."

Suits believes there is an ultimate gaming goal in every game, which is a "state": in table tennis, the goal is to get the ball over the net; in a 400-meter race, the goal is to run from the starting point to the end; in checkers, the goal is to capture all the opponent's pieces. These goals are unrelated to the game itself but refer to the state of things.

Once there is a game goal, it is necessary to have a game method to achieve this; however, the method is not always effective. For example, in table tennis or a 400-meter race, interfering with other players or jumping the gun cannot be seen as an "effective" method, or the proper way to win. And "ineffective" methods are why there is social chaos and why rules are needed. This means that we need to focus on the method that helps us to win the game. And this method, a.k.a. the game method, is the proper method to achieve the game goal.

The next question is "what limits the use of methods in the game?" The obvious answer is its rules, which prohibit certain methods from being used. In table tennis, it is effective to interfere with the opponent, but prohibited. To win a 400-meter race, one could run straight across the playground to the finish line, or fly a plane over it; the best way for Lee Sedol to defeat AlphaGo would be to unplug its power supply; one could win the game of rock-paper-scissors by showing their hand last, but it is forbidden. And these rules and the well-defined pre-game goals together construct the conditions that must be met to play a game.

This is also the greatness of human civilization for thousands of years. When we invented Go, we set the rules first, studied the gameplay, defined what is good and what is not, and made it cultural and metaphorical. Most importantly, AlphaGo came out because Go has been made interesting and meaningful to people. In another word, what matters is the ability to attach importance to rules. Like a piece of land, when people establish a border, it becomes valuable.

By observing the rules of many games, it is noted that the establishment of rules always excludes the simplest, most effective, easy, and direct methods – preferring complex and difficult ones. Therefore, people are prohibited from using the most effective methods to achieve pre-game goals, but encouraged to use "less efficient" methods to win the game. That is to say, the game is played under constraints, and achieving the ultimate game goal under such constraints requires players to explore.

With the most basic rules, sometimes other rules are extended on top of them, like those with a punitive nature. For example, the three-second defense rule of basketball. Violating it will not end the game but activate a punishment. Such rules are merely an extension of the main rules. In the game system, from the goal to the methods to achieve the goal, and to the rules that limit the methods, the logic of the entire game is built, and the game's world is constructed, which is applicable to real life as well.

In a sense, the real world is a normalized system, which is like a fork in the road. Different forks correspond to different individual goals. People move forward by making choices, with different choices and actions leading to different results. Obviously, these choices make sense. Although we are only entitled to "limited freedoms" under such regulations, it is these "limitations" that make our choices meaningful.

From Game Mechanics to Real World Rules

Both the game world and the real world follow a balanced unity of goals, rules, and methods. Different rules impose different constraints and punishments, thus deriving different game methods to achieve the game goals. In order to better balance the relationship among game goals, methods, and rules, a set of systematic mechanics is needed.

It is worth noting that mechanics and rules are different stories. The word "mechanics" comes from ancient Greek and originally refers to the structure and working principle of the machine: what components make up the machine and why; how the machine works and why. When the original meaning of mechanics is extended to different fields, different mechanics come into being. From management mechanics to social mechanics, from biological mechanics to game mechanics, whatever kind of mechanics in whatever field, it connects all parts of things in a certain way of operation, so that each part can function in coordination. From this perspective, mechanics is a balanced system based on existing content.

In gaming, mechanics is more of assistance – complementary to the rules – so that players can have a better game experience. For example, the reward and punishment mechanics supplements the presence of game rules to enhance playability. Also, the rules are open and consistent to all players, while game mechanics are not necessarily open. Perhaps most players know the game mechanics is allowed to stay hidden: for example, when a player is at a disadvantage, the game may automatically lower its difficulty to improve the player's sense of participation, proving that mechanics coordinates and balances the game experience.

At last, as the player interacts with the game, the gameplay is born. When mapped to real life, the different "gameplays" are the different lives of the people. In fact, whether it's reading books, traveling around the world, or becoming a craftsman, from the perspective of game thinking, it's all a kind of gameplay. It helps people understand the phenomena for a long time, and slowly deepens our understanding. There are endless phenomena, so there is no distinction between knowing this phenomenon and knowing that one. Aware of this fact, one can become interested in many things and broadminded.

At this point, the game mechanics have completed the transition to real world rules. This is why virtual games matter to real life: from this angle, real life is also a game. Under the dynamic balance of goals, rules, and methods, people find their own ways of survival. In such a long-term game process, more innovative mechanics may arise, breaking our perception of the boundaries of the world again and again. Yet, this is certainly not the end of life innovation, as human beings will not be satisfied with existing game mechanics. On the contrary, we will break through their constraints, look for more game elements, and bring players newer game experiences.

Before the advent of the Internet, people could only engage in limited game activities in physical space. And the leap of information technology has taken human society into a virtual space. From the arcade game era to the era of consoles, and then to the era of mobile phones, video games have begun to deliver their content interactively. Players get a variety of emotional experiences through these games. Every player has their own interpretation and opinion, and gradually grows attached to the game.

Whether it is *Plague Inc.*, which attained impressive popularity during the COVID-19 pandemic, or *Animal Crossing: New Horizon*, which was major hit in 2020, or more mobile games. After all aesthetic performance, technical realization, and story settings are stripped, people finally discovered that virtual games are not real, but approach it as they observe it. Games originate from the real world and inspire it. The idea, of the "Game of life" is actually much more profound than we think.

5.2 Presentation of the Metaverse

Thanks to technological advancements, video games have also undergone a dramatic change from crude simulations at first to now giving players the power to create worlds and play God. Now, on all kinds of robust platforms, players have the opportunity to pilot starships, harness magic, and slay monsters. Game companies, deeply aware of what players want, integrate mathematics, aesthetics, psychology, and art into their video games, trying to give them an extremely intense experience.

Consequently, as games better simulate reality and extend reality, they have also come closest to the concept of the metaverse. The gaming content production builds the content foundation of the metaverse; the technical iteration of the games supports the metaverse's technical system, and the commercialization path is the breakthrough of this pioneering industry to the metaverse.

Video Game Development

In 1946, the world's first computer – *Eniac* – was born in the US. Twelve years later, the Brookhaven National Laboratory in New York developed the world's first game

– PONG – that used a transistor as a display. Boredom is a kind of productivity, too. Since computers were invented, the interactive toys that many electronic engineers made in their spare time became video game prototypes.

Soon, it was discovered that this new type of electronic entertainment held great promise. Many were attracted to such games, so more and more people began to try to develop interesting video games, having discovered that PONG was more fun than national lab work! Eventually, Nolan Bushnell made the first commercial game – *Space War*.

Soon after, he established Atari Inc. When Atari was making a fortune, the Japanese company Nintendo also noted the huge business opportunities of the budding industry. Nintendo, which used to be a playing card manufacturer, turned to the electronic toy market thanks to Yamauchi Hiroshi's sharp business acumen. And it was Nintendo that truly brought games to millions of families.

Nintendo's innovative use of game cards as a medium perfectly integrated the relationship between software and hardware, and also laid the foundation for later game sales. The development of computer technology is particularly manifested in video games; the expansion of memory and the advancement of computer graphics can be directly seen in game images.

With technical support, the game industry was quickly embraced by the world. In the following years, more and more companies entered the game industry. Since the 16-bit computers, Sega has also joined the competition. Ans since 32-bit computers, Sony, with its PlayStation, and Microsoft, with its Xbox, embarked on 3D game road.

The emergence of PlayStation injected fresh blood into the industry. And the roaring success of the PS1 made everyone eager for its follow-up models: the PS2 was released in 2000. It is the best-selling console by far, with over 150 million units sold. In addition, the geometric growth of computer memory and graphics has enabled game developers to run wild with their imagination and dream big.

The beginning of the 21st century was the golden period for the development of client games. Classic foreign games from South Korea, Europe, and the US continued to enter the Chinese market, and domestically-developed large online games like *Fantasy Westward Journey* continued to spring up. Around 2007, with the development and application of Internet broadband and Flash, simple and convenient web games rose. The player base further expanded, and the types of games have continued to improve. In 2010, the development of client games reached its peak.

With the popularization of smartphones, the improvement of network transmission speed, and the fragmentation of user time, mobile games gradually budded in 2011, and entered a period of rapid development in 2013, driving continuous growth in China's game industry. According to the China Audio-video and Digital Publishing Association, the revenue of the Chinese game market in 2020 was RMB 278.69 billion ($43 billion), an increase of 20.71%. Of the total, mobile games accounted for RMB 209.68 billion ($32 billion).

Online games simulate real life games through science and technology, allowing people a similar experience when conditions are insufficient for real-life games. Based on this, online games have become a safe adventure. The high fidelity of image simulation and the real sense of control make every player unknowingly connect themselves with their avatars in the game. That's why good games have become works of art through the creation of fantasies.

People can live a moving story in *The Last of Us*; appreciate the watercolor paintings in *Ori and the Will of the Wisps*; feel the dynamic rhythm in *Persona 5*; or create some performance art. For example, Japanese pixel artist BAN-8KU exhibited his works in *Animal Crossing: New Horizon*.

Regarding client games, web games, and mobile games, and supported by terminal hardware upgrades, the continuous development of the industry provides users with greater game convenience, richer gameplay, and more detailed and refined content.

Further Game Extensions

Game development, as a virtual world constructed on the basis of reality simulation, shows certain similarities with the metaverse. Compared with games, the metaverse is a persistent and stable real time virtual space. It holds a large number of participants, and allows almost all behaviors in the real world to happen in the virtual space. There is also a fair closed-loop economic system in the metaverse. Meanwhile, users can continue enriching and expanding the virtual space margin through content production. Games form the underlying logic of the metaverse, while the metaverse is a further extension of games.

Primarily, there is a virtual space in both games and the metaverse. Games build a virtual world with boundaries by establishing maps and scenes. For example, the game *GTA5* creates a vast land of Los Santos for players, and provides them with great freedom of exploration through refined scenes; the AR game *Pokémon GO* creates a Pokémon world based on real-world scenes for players to explore. However, both an open-world game and an AR game based on real-world scenes are the basis for the metaverse, which also requires the creation of a virtual world with continuously expanding boundaries.

Secondly, both games and the metaverse give users a corresponding virtual avatar. They can customize the appearance, perform a series of operations such as entertainment, social interaction, and transactions based on these avatars, and build a series of social relationships. Tencent's *Moonlight Blade* is an example that enables personalized avatars. Also, games and the metaverse constitute an environment with high cognitive requirements through rich storylines, frequent interactions with players, vivid images, and proper sound effects. Thus, players must utilize a lot of brain resources to focus on what goes on in the game, resulting in the sense of immersion.

Finally, game engines are necessary for the metaverse to create highly immersive and lifelike virtual worlds. As a super-digital scenc for super-large real-time interaction, the metaverse requires a variety of capabilities to sustain its high fidelity of simulation and vast information. Therefore, game engines must continue breaking through the next-generation of technical capabilities as they develop more vivid effects.

At present, Unreal Engine 4 and Unity 3D Engine are commonly used in the game industry and have realized advanced functions such as physically based rendering, Sub-Surface Scattering, and GPU particles. Certainly, game engines are still advancing toward greater power and easier use. For example, Unreal Engine 5, as an iterative version of the engine across generations, has greatly optimized the development work-flow, and improved rendering efficiency several times for better effects. In the future, with the continuous upgrade of engine capabilities, easier-to-use engines with greater power and higher simulation fidelity are expected to accelerate metaverse development.

Start with Content

At present, there is a clear trend for games to be mobile, refined, and popular. However, as the primary form of the metaverse, one single game still has a lot of room for improvement in terms of immersion, freedom, and content derivation compared to the metaverse. Among them, content derivation holds a more critical meaning because the starting point of the metaverse is not a platform, but content that can stay independent, self-iterate, and multi-dimensionally attract users to experience or even create it.

Building a metaverse starts with content. From content to platform, it expands. It does not begin with opening a platform. A boundless universe is never born out of thin air. Therefore, in order to realize the transition from games to the metaverse, different game developers have adopted different paths, including professionally-generated content (PGC), professionally-and-user-generated content (PUGC), and user-generated content (UGC).

The PGC games are characterized by the content of the developer's self-built platform and player participation and interaction. They provide users with a highly immersive and free exploration experience. For example, the aforementioned classic game, *GTA5m* creates a vast virtual city, Los Santos, with rich details and a high degree of freedom. As the main storyline advances, players can freely explore the city in great detail, participate in a series of non-linear side missions, drive modified vehicles for street racing, and conduct a series of operations that cannot be done in the real world.

According to Take-Two's 2020 financial statement, the cumulative global sales of *GTA5* have exceeded 145 million copies. It is wildely popular among game users. In addition, other games with professionally-generated content, such as *The Witcher* series developed by CDPR, and *Genshin Impact* by miHoYo, have achieved excellent market performance and earned fine reputations. It is foreseeable that with the continuous upgrade of game engine capabilities, the metaverse will attain more efficient rendering effects and greater scene richness. It will also remove conventional map boundaries and become a completely open world.

The games with PUGC are developed in the form of a "supermarket." There are both freedom and external products. For example, *Minecraft*, developed by Mgjang Studio (acquired by Microsoft in 2014), is created with weak centralization, basic storytelling, and a high degree of freedom.

Players can explore and interact in this randomly generated 3D world by collecting ores, fighting hostile mobs, crafting new blocks, and collecting various resources available in the game. Meanwhile, it allows them to construct buildings and create art in single/multiplayer mode, and implement logic operations and remote actions through redstone circuits, minecarts, and mine tracks. The ample creative freedom of *Minecraft* has therefore attracted a large number of players. According to official data from NetEase, there are more than 400 million mobile and PC game players, and over 50,000 pieces of high-quality creative resources in *Minecraft*.

The games with UGC rely on the platforms to provide a marketplace, where all players can freely produce and trade. *Roblox* is the most representative UCG game. It provides simple and practical authoring tools to help producers generate rich and interesting game content, thus attracting users to the platform. Through personalized avatars, social interaction, and slick game content, players often immerse themselves in the Roblox platform. In recent years, Roblox has witnessed rapid development, becoming an online game platform that has swept the global youth group.

5.3 Roblox: A Wider Metaverse Boundary Based on UGC

In March 2021, Roblox, an open game creation platform, went public, becoming the "first stock of the metaverse." Its stock price rose 54% on the first day, and its market value exceeded 40 billion dollars – ten times its 4 billion valuations from a year before. Subsequently, a series of related financing strategies have occurred, making the metaverse one of the hottest concepts in the capital market. 2021 was truly the "first year of the metaverse."

Sandbox Game Leader

Roblox, based in San Mateo, California, was founded in 2004. It began with children's education and later transformed into a sandbox game company. Sandbox games, evolved from sandplay, are a unique type of game. Usually, such games have a vast map and contain a variety of game elements, including role-playing, action, shooting, etc. Creation is the core theme and objects available in the game can be used by players to create their own things, changing or influencing or creating the world. There is no main storyline in most sandbox games. They generally take player survival as the first goal, exploration and construction as the second, and changing the world as the ultimate objective.

Regarding development history, Roblox has gone through three stages successively: through updating the main body, the iteration of the development platform, and the economic system, it has achieved breakthroughs on the product side, and accelerated its commercialization and internationalization.

In the first stage from 2004 to 2012, Roblox initially built a platform and community structure. David Baszucki and Erik Cassel co-founded the company, having previously founded Knowledge Revolution, a physics simulation-experiment teaching software company. This gives Roblox the "self-built experiment" gene as well as its UGC attribute. On this basis, the company initially built a platform structure, launched its first version, Studio, and beta client in 2006. Then it released the virtual currency Robux in 2007 to build an internal economic system.

2013–2015 is the second stage, in which Roblox commenced commercialization and continued to iterate its engines. A typical sign of the company's commercialization was the introduction of the Creators Trading Program in 2013, allowing developers to obtain Robux through micro-transactions, in-game sales of virtual goods, etc. This change has connected users and developers. In addition, the company continues to iterate its engines and improve infrastructure, establishing its first data center in 2014 and strengthening its global server architecture.

From 2016 till now is the third stage, in which Roblox has accelerated commercialization and internationalization, and continues to improve its engines and community construction. Connected to Xbox and Oculus in 2016, it has effectively completed the cross-platform layout of PC, mobile phone, console, and VR devices. On this basis, it began a Premium paid membership system in 2019, and launching the Avatar virtual item trading market. It has also entered South America, Russia, and other global markets, establishing Roblox, a joint venture with Tencent.

The Roblox Platform

Roblox's product of the same name, Roblox, is a platform for online games and game creation. It was officially launched in 2006. Most of the works are generated by users themselves. In these games, players can also develop various forms and categories of games, realizing a variety of gameplay in a single game. Roblox provides simple and practical authoring tools to help content producers develop rich and fun games with UGC and attract players to the platform. And through personalized avatars, social interaction, and rich game content, players often immerse themselves in the Roblox platform.

Mainly oriented toward children and teenagers, Roblox aims to provide a suitable space for them on the network to create and play games. The early entrepreneurial experiences of David Baszucki and Erik Cassel (Roblox founders) based on game + education laid the foundation for its popularity amongst teenagers.

For the entirety of 2020, Roblox users under 9 and from 9–12 years old accounted for 25% and 29% of users, respectively; combined, those under 13 made up 54% of the users. Despite its youthful demographic, the platform continues to penetrate higher age groups. In 2020, the year-on-year growth of daily active users for groups under 13 and above 13 was 72% and 106%, respectively. The growth of the over-13 market was higher.

Roblox products include Roblox Client, a virtual experience platform shared by users, Roblox Studio, and Roblox Cloud. Roblox Client is an application that allows users to explore the 3D digital world; Roblox Studio is a toolset built for developers and creators to create, publish, and operate 3D experiences, and other content that can be accessed through Roblox Client; Roblox Cloud includes the services and infrastructure that power Roblox's platform shared by all users.

Roblox Client: an Immersive Experience

Roblox Client provides players with consistent, immersive avatars across multiple devices, creating a strong sense of substitution and identity. Roblox has a built-in Avatar Editor, which supports players to modify, design, and create the look, body, clothing, and movements of their virtual avatars. Players can also purchase specific ready avatars from the store. Avatar Editor gives players great freedom for personalization, and Roblox will match the images that the players have set in each device to ensure they are consistent in their game experience.

In addition, the Roblox platform has both traditional games such as shooting, fighting, and parkour, and many other games that are difficult to define with conventional categories. For example, in *Adopt Me!*, players can adopt children or pets, get resources to decorate houses and buy tools to take care of them, make friends with other users, and host parties; in *Robloxan High School*, players can be a royal princess, attending classes or balls, etc.

In addition to games, Roblox attaches great importance to virtual social interaction. It is not only a game platform, but a virtual social life platform, with plenty of social games. During the 2020 COVID-19 pandemic, gameplays such as *View Nearby Players*, *Online Meetings*, *Party Place*, and *Virtual Concerts* have been added to further promote the development of in-game social activities.

Currently, Roblox supports iOS, Android, PC, Mac, Xbox as well as VR devices such as Oculus Rift, HTC Vive, and Valve Index. It ensures that players only need to connect to the Internet to enter the virtual world and start interacting with other players. And its compatibility across multiple devices, consistency, and highly personalized user images give players a strong sense of substitution and identity.

Roblox Studio: Multi-platform Deployment

Roblox Studio is a platform of tools that allows game developers and virtual item creators to build, publish and operate 3D experiences and other content. It is both low-cost and easy-to-use, and the platform community strongly supports developers, effectively lowering the threshold for game development.

Roblox Studio is an open game engine. Game engine refers to the core components of some pre-written editable computer game systems or some interactive real-time graphics applications. These systems supply game designers with various tools needed to write games so that designers can easily and quickly create game programs without starting from scratch. The game engine includes the following systems: rendering engine (including 2D image engine and 3D image engine), physics engine, collision detection system, sound effects, script engine, computer animation, AI, network engine, scene management, etc.

Roblox uses the RBX.lua language, which is completely free to learn. Lua is both the front-end and back-end development language of Roblox, and the learning cost is fairly low. The engine itself holds many functions, such as backpack, chat, team, and other systems, which developers can access directly. Roblox also has a complete framework on both the client and the server, so that developers can immediately start developing without knowing the complex game framework. Compared with mainstream game engines such as Unity and Unreal on the market, Roblox has relatively inferior image quality for now; however, its game development is completely free and highly flexible, while its development language is less difficult to learn.

Roblox Studio supports developers with a developer center, beginner's tutorials, community forums, educator centers, and data analysis tools. And the developer center (Dev Hub) is equipped with an API library, tutorial collection, and other practical materials. The community forum is a communication platform for developers to discuss new platform features, community events, job opportunities, bug reporting, and have direct communication with Roblox employees. The educator center (Edu Hub) offers guidance on programming tutorials, 3D designs, and community rules for teachers, students, and parents who are learning to program. All Roblox developers have a dashboard that shows their daily traffic and Robux (the platform's virtual currency).

In addition to game creativity and development, Roblox Studio also renders developers with a full range of services such as distribution and channels. Different from traditional game engines, it enables developers to focus solely on game creativity and development: for it renders services such as data backend, operation, and maintenance disaster recovery, instant messaging with friends, network communication, and game distribution channels.

The vast developer ecosystem feeds Roblox with diversified gameplay and different themed games, satisfying user needs to try several games. In 2020, Roblox platform users opened 20 games every month on average, and they tried more than 13 million games in total. Judging from the game categories with the highest turnover on the platform, its players have a high degree of acceptance of different gameplay and themes.

Among its 15 games with the highest turnover, there are diverse categories: not only mainstream types, such as MMO and FPS, but also the more obscure, such as pet socializing and parkour. Regarding themes, there are not only campus and pet themes for the platform's child players, but animation themes for older players.

Roblox Cloud: Easy Play

There is a cloud architecture based on its own infrastructure on Roblox. Most of the services operated by the Roblox Cloud are entrusted to self-managed data centers, while for some high-speed databases, scalable object storage, and message queuing services are used. When additional computing resources are required, Amazon Web Services are employed. All servers responsible for simulating virtual environments and transferring materials for Roblox clients are private possessions of Roblox, and are widely distributed in data centers across 21 cities in North America, Asia, and Europe. And such wide distribution gives Roblox strong disaster recovery capabilities.

According to Portworx, Roblox Cloud adopts a hybrid cloud architecture based on its own infrastructure. And Roblox's private data centers and edge computing nodes connect users and some AWS external cloud services. Roblox Cloud is expected to achieve the same global server, allowing customers to quickly start the experience on different devices. After the user clicks on the game, the client will immediately start simulating and rendering the virtual world with lower quality detail. And the simulation quality slowly increases as the user receives more materials with better information.

Roblox Cloud transmits materials, and adjusts their formats, details, and priorities to optimize the bandwidth and functionality available through a network of nodes distributed in different regions. And all simulations of virtual environments are performed on Roblox private servers. The company's prospectus discloses that it currently supports millions of players online simultaneously. With the further improvement of server capabilities in the future, it is expected to achieve the same global server.

First Metaverse Stock

On March 10, 2021, Roblox successfully released its IPO on the New York Stock Exchange through DPO; its stock price closed with a 54% surge on its first day. Based on game freedom and excellent user activity, Roblox appears to be the metaverse prototype for the time being. It is the first company to include the metaverse in a prospectus; this new element has ignited great interest in tech and investment circles as well as the infinite imagination of players.

Primarily, the Roblox developer community is active, and creators are fully motivated. As of 2020, a total of 1.27 million developers have profited from developing games on Roblox, and players have opened 8.45 million games. Among them, three developers have earned over $10 million, and 272 games have been on the market for more than 10 million hours. The leadership effect has begun to show. Meanwhile, Roblox can still meet the long-tail needs of players. The total participation time of the Top1000-Top50 games accounted for 34% of the total, while the long-tail needs outside the Top1000 occupied 10%. Roblox games are rich in content, and such content is updated and iterated actively. The company's prospectus disclosed that among the top-ranked games from 2015-2020, half were produced in the previous two years, and one-third were in the last year alone.

Creators can profit by selling experiences (games) and in-app purchases on the platform, earning creator rewards based on user engagement, selling development tools and content to other creators, selling goods on the virtual item market, etc. Their income will remain on their virtual accounts, and developers who meet certain criteria will be able to convert that into US dollars through the Developer Exchange Program. In 2020, a total of 4,300 developers earned $329 million through the program.

Secondly, there are a large number of social games on Roblox, thus a strong community vibe. Most Roblox games are relatively easy, with low thresholds for playing. Common users can play them on multiple devices such as mobile phones and computers. And these games emphasize real-time interaction with online and offline friends, and exude a strong social leisure vibe.

According to Bloxbunny statistics, three of the top five most played games in the 30 days ending June 15, 2021 were social MMO games. The No.1, *BrookhavenRP*, had 395,700 users online at the same time. *Roblox* itself is strongly social, too. Players have a consistent avatar in each game, and can befriend each other and chat. Meanwhile, new social forms such as "Play together" and "PartyPlace" are opened to enrich the platform's social experience.

Lastly, *Roblox* has witnessed a flywheel effect from an active developer ecosystem and an active user ecosystem. The company is committed to building a high-quality UGC game platform. The active developer community makes it more attractive and stickier for users. That more users play and consume on the platform generates abundant income for content creators and catalyzes more active developers – the flywheel effect. In addition, while the platform is strongly social, its users are highly active, too. It maintains low customer acquisition costs and strong community stickiness through its strong social relationships.

Since 2019, the company's marketing expenses have exhibited a downward trend, while the number of daily active users has maintained rapid growth year-on-year, indicating that its customer acquisition does not necessarily depend on advertising and marketing, but more on word-of-mouth and natural social growth. According to

Broadband Search, the daily active usage time of major social media is 20-60 minutes, while that of *Roblox* in 2021 Q1 reached 153 minutes.

5.4 *Axie Infinity*: Build the Metaverse Economy

Axie Infinity is a decentralized turn-based strategy game based on the Ethereum blockchain. Players can raise, fight, breed, and trade digital pets of the NFT elf, Axies. Blockchain games turn the digital assets in the game into NFTs. As blockchain cannot be tampered with and records can be traced, property rights are recorded, and the authenticity and uniqueness are ensured, game asset transactions no longer rely on the company platform for security guarantees. Each elf Axie in *Axie Infinity* is a unique NFT, and the ownership and transaction records are publicly displayed on the blockchain.

By building a complete player entry and exit mode, and designing AXS coins, exchanges, and community treasury outside the game, *Axie Infinity* has become the first game built on NFT that generates a large amount of revenue.

ETH, SLP, AXS

There are three core NFT tokens in the *Axie Infinity* ecosystem: ETH, SLP, and AXS.

ETH: it is the only point of interaction between the game ecosystem and the external economic system. To play, there must be a team that holds 3 Axie NFTs (at the current market price, the minimum entry fee is about 4000 yuan). Axie can only be obtained through trading, hatching, and gifting. To play the games, new players must make the purchase with ETH. Also, sellers who successfully make an Axie deal will bear 4.25% of ETH as handling fees.

SLP: as the main profit model of Play-to-Earn, SLP is obtained through player-versus-environment (PVE) or player-versus-player (PVP), with an upper limit of 75 SLPs per account per day (up to 50 in PVE, 25 SLPs after 5 PVP victories and 10 PVE victories). The SLP acquired by players is mainly used to breed a new generation of Axie NFTs or is directly sold. And obtaining SLP through in-game battles or daily tasks, and accumulating it to achieve the purchase value of Axie NFT is the most important profit model in the game.

AXS: as an official governance token, AXS can only be obtained through monthly ranking rewards, and only breeding in the game consumes it. Moreover, AXS holders can participate in the platform's governance voting by pledging AXS. The *AXS Universe* consists of players, AXS holders, Axie's Game Universe and Axie Community Treasury.

AXS, as a digital currency issued by Sky Mavis, represents the entire *Axie Universe*, and constitutes the most basic goods for interaction between players in the *Axie Universe*. In the game, players get AXS through trading and playing games; they spend it through gaming and raising game pets. Outside the game, players who hold AXS can invest in the Axie Community Treasury and receive voting rights for game decisions, and dividends provided by the game as per their shares. Therefore, under this economic system, AXS forms a complete closed economic loop. And game revenue is no longer monopolized by developers: the total circulation of AXS is 270 million, and there will never be more. Of this, 20% is used for playing the game and profiting, while 29% is used for pledging and rewarding.

Build a Balanced *Axie Universe*

The breeding system is an important node for the *Axie Universe* to complete the closed loop. Players can breed a new Axie by pairing two Axies, whose genes determine the attributes of the baby Axie. However, there are only seven breeding opportunities per

Axie, which is a factor that affects the sale price. Each Axie has four basic attributes, and the differences in their attributes determine their combat stats. In addition, each Axie consists of six parts, which determine what skill can be used. These parts are controlled by three genes, namely, the dominant gene D, the recessive gene R1, and the sub-recessive gene R2.

The complex genetic system combined with the limited number of hereditary options and its randomness greatly increase the difficulty of breeding a perfect Axie. And the complex combination of attributes and skills makes each Axie distinct, giving players room for continuous improvement. When the breeding opportunities of an Axie are exhausted, the resources invested in it go to waste.

In the breeding system, the emergence of the SLP token actually forms a closed loop. There are both output and consumption channels for SLP – it is tradeable. This is a way for common users to make some money, because the birth of any new Axie consumes SLP, and the more reproductions, the greater the consumption. For Axie with better genes, the consumption during later reproduction will be considerable. The continuous birth of new Axies provides a wealth of optional Axies for new players, and Axies with excellent attributes are often not cheap. This fuels the constant birth of new Axies, as well as the constant consumption of SLP.

In addition, the combat mechanics adds game diversity to the *Axie Universe*. It is the main profit scenario for players. The game's most important income source is the daily PVE/PVP tasks. Players get SLP rewards through Axie battles, which they then invest in Axie breeding. The game data in Axie Infinity is relatively balanced: the battle deck consists of three Axie's 12 skill cards. The hand starts with 6 cards and 3 energy points, and each round, players draw 3 cards and increase energy points by 2.

At the same time, it is also important to judge what special skills should be used according to different combat situations. The large number of cards drawn in each round and the forbidding of discarding cards minimizes randomness, and tests the rationality and pertinence of the player's skill configuration. Compared with the previous blockchain games that solely focused on the transaction attribute, Axie Infinity has significantly improved the playability. The trading targets are no longer limited to players who resell for profit, but include those who truly want to collect powerful Axies for battle, which further enhances the value-proofness of assets and the vitality of the trading market.

Official statistics show that the Axie Infinity market has witnessed 1.7 million transactions in 30 days, and the total transaction volume has reached 1.04 billion dollars, far exceeding all other games of the same kind. And one Axie was sold for 611 dollars on average. During transactions, Axie Infinity collects a 4.25% handling fee, which is the second largest source of game revenue. Axies are directly purchased on the official market. And the total number of Axies available for the time being is 229,443. Customers can select different types of Axies to purchase through the filter box on the left. And different types of Axies have different characteristics and need to be reasonably configured for battles.

No.1 Blockchain Game

In the balanced *Axie Universe*, Axie Infinity holds a steady No.1 among all chain games, crushing the traditional frontrunners. And its monthly revenue is advancing triumphantly: AxieWorld data shows that *Axie Infinity's* revenue in August 2021 reached $364 million, an increase of over 85% from $196 million in July. Its August revenue was only second to Ethereum, which realized a total of $670 million – ranking top in chain games. Meanwhile, its success also marked a breakthrough in the commercialization of blockchain in the game field, and its monthly revenue has far exceeded that of *Honor of Kings*, which was $231 million (in July).

Axie Infinity directly links developers and players. 95.75% of the revenue is distributed to community players in the form of tokens, and every player can achieve

Play-to-Earn. And as the game develops, players can enjoy the development dividend (community treasury), while voting via holding tokens (DOA community governance). This series of mechanical innovations has greatly encouraged traditional game players to enter blockchain games. As per the Axie Infinity Twitter account, its daily active users have exceeded one million.

In addition to the game business, Sky Mavis is running three more categories of principal business: blockchain game incubator, digital currency wallet, and Ethereum's side chain Ronin.

Lunacia SDK, a player development tool that Sky Mavis promised to launch in 2022, will initially serve as a map editor that players can use to create games and other experiences. From a long-term perspective, this is an important addition to the *Axie Universe* as it enables players to be future content producers, and gain full interaction within the universe.

Also, Sky Mavis has been trying to expand ecological cooperation for Axie to gain greater popularity. In the design of the game's economic system, it cooperated with the DeFi project, and incorporated stable tokens directly linked to physical currency into the game ecosystem, which has greatly improved the user experience. And its cooperation with Samsung Blockchain Wallet in 2020 has given the project greater exposure.

At present, the overall ecological environment launched by Axie Infinity encompasses player voting rights, community treasury games, Lunacia SDK, etc. According to the plan envisaged in the white paper by Sky Mavis, by 2023, the development team will lose the absolute right to vote on *Axie Infinity*. At that time, the game will be dominated by players who hold AXS tokens and complete decentralization will take place. If everything goes as planned, this will be the first game in history to be fully decentralized and realize a self-sufficient ecosystem.

5.5 *Fortnite*: To the Metaverse

At present, though the Internet is built on open and universal standards, most giants such as Google, Facebook, and Amazon resist data crossover and information sharing, hoping to establish their own barriers and retain users. However, this goes against the requirement to communicate with the metaverse platform and share content.

Clearly, one of the most important elements in creating a shared universe is interconnection. Like currencies that can be exchanged between countries, what users buy or create on one platform needs to be seamlessly transferable to another and available for use. *Fortnite*, as a battle royale, has successfully achieved this to some extent.

A Whole New Social Space

In fact, as early as 2011, Epic Games showcased *Fortnite* at the Video Game Awards. The game was developed by the team of Epic producers, Ciffy B and Lee Perry. It is a PVE game that involves destruction, construction, and shooting to demonstrate the power of the company's Unreal Engine 4 launched the same year. In the game, players must master construction skills, and collect high-quality building materials from all over the world during the day to constantly optimize their fortresses to resist and repel the waves of enemies that attack when night falls.

Six years later, on July 21, 2017, *Fortnite* was officially released. In addition to the PC version, it was also made available on PS4 and Xbox One. With prices ranging from $39.99 to $149.99, there are multiple editions, including standard, deluxe, super deluxe, and limited. As a result, it has gone viral in the West with its comprehensive gameplay and accurate grasp of young people's preferences. In the official data released in May 2020, over 350 million registered players were confirmed.

From the perspective of the metaverse, on the one hand, *Fortnite* has unique platform interoperability and content sharing; on the other hand, it realizes the interaction between the virtual world and the real world. Epic Games has persuaded major game platforms to allow *Fortnite* to operate across platforms. The rules, combat features, and graphics styles in every version are the same. Mobile game players can play together with PC or console players, and players can access existing skins or items from other editions when they log onto a new platform.

Another highlight of *Fortnite* is the simultaneous launch of various real-life IPs, thereby further blurring the boundaries between gaming and reality. Like the American drama *Once Upon A Time*, each fairy tale is not separate. Instead, Snow White, Aladdin, Cinderella, etc., live in the same town with intertwined storylines.

Outside of games, *Fortnite* has evolved into a social space, blending game life and real life. It becomes an autonomous game. In simple terms, just like in people's daily life, when they have a free afternoon and can get together, they are not predestined to play a certain game (not only a video game), and even if they do, they are not bound by specific rules. *Fortnite* provides a free and open venue for people.

Guardian journalist and best-selling author, Keith Stuart, when talking about *Fortnite*, likened it to children's skate parks in the late 1970s and early 1980s, "There are manors, malls, factories, farms, and lots of open countryside space. Sun shines through the woods and butterflies are dancing. One can choose to form a squad of four and work together to stay alive. And since people spend most of their time exploring and looting houses to find useful items and weapons, they get "pause time" to chat. Conversation often strays from the game itself, so Fortnite is like a skatepark – a social space and a playground."

People can wander or hop around the island, or explore from the ghost church tower in Ghost Hill to the labyrinthine tunnels under Shifty Shafts (a location on the *Fortnite* map), or dance at the discotheque. There is even a soccer field in the middle of Pleasant Park! It provides a safe place for people to roam around and a whole new place to socialize online.

As a competitive game that up to 100 players can play simultaneously, *Fortnite* is different from other competitive games. Socializing is a core element: in the face of tasks, players need to cooperate with others through different organizational forms, such as guilds and teams, to overcome difficulties. In the player community, each player has to sell various forms of labor to obtain the items they need, which forms a relatively primitive market. In a virtual social group, players connected by the game use knowledge of specific fields, exchange skills, experience, and resources to compete and cooperate with each other.

Fortnite is believed to be the closest system to the metaverse at present. It is not entirely a game, having evolved into a social space where people use virtual avatars to interact. As of April 2020, the total game time of 350 million registered players exceeded 3.2 billion hours – the longest game time (online time) of any game in the world.

In February 2019, Marshmallow held *Fortnite's* first live concert. In April 2019, Avengers: Endgame introduced a new game mode where players took on the roles of the *Avengers* and fought Thanos. In December 2019, *Star Wars: The Rise of Skywalker* held a promotional meeting in the game, and director J. J. Abrams was interviewed live. In April 2020, American rapper Travis Scott staged an immersive concert called Astronomical on major servers around the world. It was watched by 17 million people at the same time and ignited a frenzy on social media. In addition to entertainment, the economic activities in *Fortnite* are more active players can create digital clothing or emojis to sell for profit, write their own games or plots, and invite others to play.

Build a Digital Ecosystem

Epic Games, the game developer of *Fortnite*, is an American-based interactive entertainment company and 3D engine technology provider. In addition to the huge success of *Fortnite*, it has also developed the Unreal Engine, which is widely used in various industries such as games, film and television, architecture, and automobiles. It has also launched content distribution systems such as the Epic Games Store and Epic Online Services. Together with Unreal Engine, they have built a digital ecosystem for developers and creators.

Unreal Engine: A Lower Threshold for Developers

Unreal Engine originated from Epic's early self-made game *Unreal*. Since its release to developers in 1999, Unreal Engine has been iterated and upgraded for 20 years and has become the core commercial engine of physically based rendering. In particular, it is a technical leader in providing 3D realistic-style digital image effects.

During game development, there are some more general requirements, such as general rendering, terrain, model import and generation, animation, debugging, and construction. Some early developers abstracted these technologies into general functions and launched them together, becoming the earliest engines. As technology continued to advance, the engine functions continuously improved, and so have their supporting development tools and usability. Gradually, they have become the underlying foundation of the game industry.

Compared with other engines, the advantages of Unreal Engine are mainly reflected in the creation of high-quality, 3D, and realistic content. Creators can use it to create, modify, and render photorealistic 3D effects in real time. It has attracted wide attention from the industry in terms of technical depth and cutting-edge technology exploration. In particular, its accumulation in industrialization and automated production is a stable guarantee for the efficient creation of high-quality content in the future.

Epic released a preview video of the Unreal Engine 5 in 2020 and introduced its new features, which immediately drew great attention from the game industry. Epic made it clear that the ultimate goal of Unreal Engine 5 is to enable creators in all industries to produce real-time content and experiences: firstly, to improve engine performance enough to create the visual expressiveness that a next generation engine should have; secondly, to improve the iterative effect, so that creators can easily select changes made in editing tools to various target device organization platforms; thirdly, to lower the threshold, and help small teams and individuals make high-quality content by providing richer and more complete tools.

In May 2021, Unreal Engine 5 released a preemptive experience version, mainly for the game industry. On April 6, 2022, in the live broadcast of state of unreal, Tim Sweeney, founder and CEO of Epic Games, announced the official release and open download of Unreal Engine 5.

Epic Online Services: Cross-Platform Gaming

In 2019, Epic Games began to render Epic Online Services. It opened this infrastructure and its own account system to the outside world for free, allowing external developers to use and build their own multiplayer online games. When developers are connected, it supports multiple account login, chat, achievement, matching, cross-platform online, cross-platform data exchange, and other functions on the entire platform; it is open to all game engines to access.

This means that external developers get free access to Epic's massive user base, including the login system, friend system, achievements, and leaderboards. Using Epic Online Services allows games to run cross-platform regardless of platform differences. Moreover, this strongly supports small and medium game developers, saving them a lot of work when there are relevant needs.

Epic Games Store: A Bridge Between Users and Gaming Companies

At the end of 2018, the Epic Game Store was launched on the Windows platform. Players can purchase games in the store as well as pay for in-app purchases. It appeals to players because of its rich content, some PC exclusive games, and its annual Mega Sale, which hands out some paid games or in-app purchases free, and offers a series of discounts.

The attractiveness of the Epic Games Store to publishers is its relatively lower sharing ratio, which is below the industry average: the world's more mainstream game distribution stores such as Steam, App Store, and Google Play, charge a 30% cut, while the Epic Game Store only charges 12%. And if the product is made with Unreal Engine, the cut for the store will be used to offset the cut for the engine.

The Epic Games Store, by working with developers, give them freebies to earn more customers and feedback. Compared with other platforms, it is backed by the entire Epic ecosystem, and as a part of the entire ecosystem, it also hopes to build a positive cycle for the entire industry.

CHAPTER 6

Who's Ready Player One?

Today, when the dividend of mobile Internet users has peaked, it is inevitable that a new generation of interactive carriers will be sought after radio and television, PC Internet, and mobile Internet.

Against this backdrop, the metaverse, as a 3D virtual space with link perception and sharing features based on the future Internet through augmented reality, stands in the center of spotlight. The emergence of such a concept gives people hope for the "next-generation Internet," and is on the cusp of current industries, giants are scrambling to enter the metaverse to gain a head start.

6.1 Fake Cusp or Real Future?

In March 2021, Roblox entered the capital market – a landmark event in the boom of the metaverse industry. It immediately set off a craze, and capital poured in. Soon in April, Epic Games, received a new round of financing of one billion dollars – the highest financing in the metaverse field since 2021. In China, in October 2020, investors valued the VR studio Recreate Games, which developed the demo of the popular Chinese independent game, *Party Animals*, at hundreds of millions of RMB based on the concept of the metaverse – doubling its value.

At the same time, major Internet giants entered the market with large bargaining chips, and many listed companies announced that they had formulated plans on interactive platforms, such as Facebook, ByteDance, Tencent, NetEase, and Baidu. What made the metaverse's virtual world that first appeared 29 years ago suddenly go viral in the market? Is the concept of the metaverse a fake cusp or the real future?

Capital Seeks a New Outlet

The development of information technology has taken human society from the physical world to the digital world. In the 1990s, the digital revolution was in the ascendance, igniting the first digital economy boom. During that period, digital technology was commercially applied widely in the consumer field. The end users of major business models such as portal websites, online videos, online music, and e-commerce are almost all consumers. This stage is also called the "consumer Internet," and "consumer platform."

However, with the development and popularization of the Internet and the development of user habits, the growth of mobile Internet users slowed, and the consumption of Internet dividend is gradually shrinking. In 2020, the COVID-19 pandemic further encouraged users online, and the user time length may slowly peak in the short term. According to QuestMobile, from 2015 to 2020, the average monthly Internet usage time of Chinese netizens continued to increase, and this number in April of 2020 reached 144.6 hours – 54.8 hours more than it was in 2015; from January

2018 to June 2020, the time that users spent on short videos has witnessed a steady climb.

In this context, the metaverse has become a new outlet for capital due to the important investment opportunities it carries. For example, in terms of 5G cloud games, the improvement of cloud computing propels cloud games to enter the warm-up stage. 5G will make up for the shortcomings of data transmission, promote the overall development of cloud games, drive the increasing popularity of consumer entertainment, and break the restrictions of time and space on media Internet services.

At present, both Chinese and foreign Internet giants, as well as gaming companies, are continuing to deepen their involvement in cloud games. China cuts in from the mobile cloud game platform, while foreign countries focus on cross-platform opportunities. The expansion of cloud game platforms will cultivate user habits to pay for subscriptions. They may further encourage the game industry to position content as the focal point. The Game Publishing Committee of China Audio-video and Digital Publishing Association estimates that the number of cloud game users in China is expected to reach 440 million by 2030; the market penetration of cloud games among all game users in 2030 is expected to be 54.3%; the scale of China's cloud game market is expected to grow to 39.53 billion RMB in 2030; the 2020–2030 CAGR is expected to reach 44.6%.

For another example, VR hardware and software are maturing day by day, business scenarios are gradually coming true, and the depth of content is constantly increasing. Since the second half of 2019, with the improvement of the VR content ecosystem and the continuous iteration of technology, Oculus products represented by Facebook have been widely acclaimed among users. Tech behemoths have scrambled to invest in AR/VR, ushing the industry into a period of rapid development.

The launch of Facebook Quest2 received wide praise in the market. The Oculus series is one of the most important products in the VR field. Compared with the first-generation Oculus Quest, Quest 2 is lighter and thinner. The price starts at $299, which is 100 dollars cheaper than the last generation. Such cost-effective products have naturally been welcomed by the market. Zuckerberg believes that only when the active users of VR reach 10 million can the VR ecosystem obtain sufficient benefits. Undoubtedly, the application demand for VR glasses will continue to rise with the development of the metaverse.

Users Expect New Experiences

Humans do have a need for a virtual world. As Milan Kundera put it, "We can never know what to want, because, living only one life, we can neither compare it with our previous lives nor perfect it in our lives to come. There is no means of testing which decision is better, because there is no basis for comparison. We live everything as it comes, without warning."

Fiction has always been the base impulse of human civilization. The ancient Greek bards sang stories of their heroes; fairies, ghosts, geniuses, and beauties were written in poetry and novels; and Shakespeare designed the witches who gently stir the poison for Macbeth in his play. The stories in the films and television dramas are the lives of others presented to the audience. With the popularity of video games, people can play a role, or become another role in an interactive way, on mobile phones and computers.

What is absent in the real world is made up for in the virtual world; when possible, there will be compensation from the virtual world in the real world. Based on such a "compensation theory," the simulation theory believed by Nick Bostrom, Elon Musk, etc., assumes that a civilization's impulse to create a virtual world for compensation is eternal. Over a long period of development, virtual worlds will inevitably be created, and the worlds in which they are located are most likely created by upper-level designers.

Jean Baudrillard distinguishes three stages in the history of human simulation: the first stage is counterfeit, which believes that value exists only in the real world, and fictional activities must simulate, replicate, and reflect nature. And truth must be distinct from its counterfeits; the second stage is production, in which value is governed by the laws of the market, with the aim of making a profit, and mass-produced counterfeits and real copies become equal; the third stage is a simulation, where the simulacra creates the "super-reality" and assimilates the reality into itself, and the boundary between the two disappears. The reality, as the mock object no longer exists, is that the counterfeit becomes a copy of what is without an origin, and the illusion is confused with reality.

The metaverse is the simulation of the third stage. It provides people with a virtual world where they can live another life. In this world, there is a complete operating world system, and people carry out daily activities in a variety of scenarios. In addition to games, they can conduct social, shopping, academic, leisure, and even exercise activities in the metaverse through external devices such as treadmills. When the metaverse comes true, people's desire to experience a wider life will have an ultimate form of realization – to immerse themselves in another world with new identities.

Technology Desires a New Revolution

Human society has undergone agricultural and industrial civilization, and eventually welcomed the digital civilization in the 21st century.

During the agricultural civilization, human society only relied on substances directly available and consumable in nature, such as plants and animals. The habitat of the Hominids was either a natural place, or one that could shelter them from wind and rain after some simple improvements, such as caves and straw sheds.

Agricultural civilization was also an age of physical strength. It only witnessed the shift from pure human labor to the use of farming tools, such as grinding stones into sharp or blunt stone axes. Ape-men used them as a "universal" tool to attack beasts, sharpen sticks, or dig up plant roots.

In the Mesolithic Age, stone tools evolved into inlaid equipment – wooden or bone handles were attached to a stone axe; The transformation from tools of a single material to those of two heterogeneous materials occurred. On this basis, human beings have developed compound tools such as stone knives, spears, and chains. In the Neolithic Age, humans learned to drill holes in stone tools and invented stone sickles, shovels, hoes, and mortars for processing grain.

The Industrial Revolution was the starting point of industrial civilization as well as a fundamental change in human production. During this period, humans experienced the development from manual manufacturing to machine manufacturing. The development of industry has enabled us to have greater power to transform nature and obtain resources, and the products it produces are directly or indirectly consumed by people. As a result, the quality of life has been greatly improved.

Since the First Industrial Revolution, industry has determined the survival and development of humanity to a certain extent. In fact, the rapid development of industrial productivity in capitalist societies precisely enabled the bourgeoisie to create more productive forces in less than 100 years than all the productive forces that had ever been created before. But whether it is an agricultural civilization or an industrial civilization, it is the people who create value, and use tools or machinery for production.

After passing through the agricultural and industrial eras, digital technology, represented by the Internet and AI, is forming huge industrial capabilities and markets rapidly, thereby elevating the entire production system to a new level, catapulting society into the digital age. Unlike previous eras, the digital civilization is a deep amalgamation of cyber-physical systems, and innovative manufacturing technologies and modes.

On the one hand, the digital era is the age of computing power. The main body of productivity has undergone a qualitative change – machines can create productive value, and core labor is replaced by AI. However, this is premised on AI developing to an adequate level of intelligence. Against this backdrop, a simulated virtual world of all dimensions, or a parallel universe may become a turning point for the efficiency and cost of AI training.

On the other hand, unlike the limited resources of previous eras, they are unlimited in the digital world, and their marginal cost is close to zero. But this requires the right environment for interaction and communication between people and people, people and machines, and machines and machines. Only in this way can computing power continue to create productive value, and this environment must connect virtuality and reality.

Therefore, both the iteration of AI and the underlying data/information interaction ecosystem verify the metaverse's inevitability. The metaverse is a new digital civilization built for the digital existence of human beings in the future. It is not only VR/AR and the complete reality of the Internet, but the way of life of mankind after 2040.

In retrospect of the past 20 years, the Internet has profoundly changed the daily life and economic structure of society; looking forward to the next 20 years, the metaverse will have a far-reaching impact and reshape the digital economic system.

Connecting the real world and the virtual world, the metaverse, as the carrier of human digital survival and migration, improves experience and efficiency, and extends human creativity and possibility. The digital world has gradually changed from the reproduction and simulation of the physical world to the extension and expansion of it. Also, the production and consumption of digital assets, and the evolution and optimization of digital twins will significantly affect the physical world.

6.2 Size of the Metaverse Market

As the market becomes heated, capital has flocked into the metaverse. A nascent market as it is, the metaverse has attracted giants to fight over it.

When the metaverse is analyzed inward, its industrial chain can be divided into seven layers.

1) Experience layer. Games are now the closest "entrance" to the metaverse, and the experience will continue to evolve from them. It provides users with more content for entertainment, social interaction, consumption, learning and business work, and covers various life scenarios.

2) Discovery layer. It is the channel for understanding the experience layer and trying to reach new users. It includes the advertising system, the process of evaluating and selecting new experiences, AND platforms like Steam, Epic Games, TapTap, Stadia, etc.

3) Creator economic layer. The experience and content in the metaverse need to be continuously updated, and the threshold for creation is constantly lowered. Also, development tools, material stores, automated workflows, and monetization methods are provided to help with creation and monetization.

4) Spatial computing layer. It seamlessly mixes the digital world and the real world, thus encouraging the two worlds can to interact and understand one another. And it includes 3D integration, VR/AR/XR, voice and gesture recognition, spatial mapping, digital twins, etc.

5) Decentralization layer. The economic prosperity of the metaverse needs to be based on a set of shared, widely recognized standards and protocols to promote unity and fluidity of the virtual economic system. Cryptocurrencies and NFTs can provide digital ownership and verifiability to the metaverse, and breakthroughs in blockchain, edge computing, and AI will further achieve decentralization.

6) Man-machine interaction layer. With the realization of miniaturized sensors, embedded AI, and low-latency edge computing systems, it is expected that future man-machine interaction devices will carry more applications and experiences in the metaverse. Because they can give better immersion, VR/AR HMDs are generally considered to be the main terminals for entering the metaverse space, in addition to wearable devices, brain-computer interfaces, and other devices that further enhance immersion.

7) Infrastructure layer. The explosion of the metaverse concept is the sure outcome of marginal improvement in infrastructure technology. With the maturity of technologies such as 5G, cloud computing, and semiconductors, the real-time communication capability in the virtual environment is greatly improved, thus supporting a sea of users to be online simultaneously. Such technology has also become key to driving increases in data volume/sophistication.

Among these, the experience and the discovery layers can be combined as the ecosystem layer, which aims to create the scene content of the metaverse; the creator layer and the decentralization layer make up the underlying structure, which provides the foundation for its economic system; the spatial computing, the man-machine interaction, and the infrastructure layers are the technical guarantees of the metaverse. Based on this, it is expected to breed a new trillion-level ecosystem blueprint.

Create Scene Content

Judging from the experience and discovery layers, the VR interaction scene is evolving from the basic and the supplementary application stages to the pan-industry application stage and the ecological construction stage.

Specifically, the basic application stage focuses on games, short videos, military training, etc. There is relatively limited content, the interaction method is monotonous, and the penetration rate in the C-end market is low; in the supplementary application stage, VR technology and content are applied to various panoramic scenarios, extending to education, marketing, vocational training, experience galleries, tourism, real estate, etc. They have just penetrated the C-end market; in the pan-industry application stage, the value of VR in medical, industrial processing, architectural design and other scenarios begins to show, and the C-end market will be expanded through B-end users. The ecosystem construction stage and VR are characterized by strong interaction and in-depth penetration; panoramic social networking will become one of the ultimate application forms of VR.

In fact, the interactive scene is the metaverse itself. Its industrial ecosystem potential is undoubtedly huge. In the long run, various application scenes will form a complete and circular ecosystem among comprehensive service, equipment, and content providers, brand advertisers, operators, and B- and C-end customers. Based on this structure, industry chain manufacturers can profit through revenue sharing, commissions, royalties, advertising charges, and other channels to maintain continuous operations.

The metaverse content scene starts with games but is not limited to them. In the future, a large number of other vertical scenes will be included, such as industrial, medical care, smart education, virtual entertainment, holographic conferences, military simulations, and social networking scenes in other vertical industries.

Taking consumption scenes as an example, with the iterative upgrade of technology, online shopping has become more visual and clearer, and the information accessible continues to be enriched. From early telephone shopping to the traditional image-text model on Taobao, to e-commerce, such as live streaming sales and social media promotion, online consumption experiences keeps upgrading. Customers used to choose the products they are interested in by comparing the pictures and descriptions; however, videos and live streaming now show users the product parameters, giving them complete product information.

From the perspective of communication, the transmission capacity of short and medium videos and live streaming is much higher than that of graphic communications. Meanwhile, with the rise of content e-commerce, a series of influencers have sprung up on Little Red Book, Douyin (TikTok), Kuaishou, Bilibili, and other platforms that share fine quality products. They provide users with greater visual-audio product information and user effects – the consumption process is reshaped. Such online platforms truly stimulate a consumer's desire to buy.

In the metaverse age, user consumption experience may undergo a new wave of interactive upgrades. Driven by innovative technology, more immersive consumption may become the norm; through the application of AR and VR, users will receive a more immersive and better shopping experience. For example, SoYoung renders users with AR facial feature detection services. The App scans the user's face through a mobile phone camera, and analyzes the proper makeup, hairstyle, and skincare products suitable for the user. Thus, customers can obtain professional beauty advice on their mobiles remotely.

The AR virtual shoe trying on the Dewu App enables users to select their favorite shoes in their preferred colors and try them on to see how they look. This helps avoid the need to visit offline shops or go to the trouble of returning the shoes when they realize they do not fit after being delivered. In the future, immersive consumption will not be limited to small items such as clothes and shoes. AR house decoration, remote real estate showing, and even simulated tourist attractions will become popular lifestyles. In addition, the amount of information consumers can reach will further increase. With the support of wearable devices and tactile sensors, there may be a better and more immersive shopping experiences than currently available.

Open up the Economic System

The metaverse is an immersive virtual world that resembles reality, so it is crucial to build it a corresponding economic system. In fact, the previous ordinary virtual worlds (online games, online communities, etc.) have always been regarded as common recreational instruments rather than real "parallel worlds." An important reason is that assets in such virtual worlds cannot smoothly circulate into the real world. Even if players put all their energy into becoming a "winner" in such virtual worlds, it is highly unlikely for their status in reality to change. Their fate in this type of virtual world lies not in their own hands. Once the operator shuts down such a "virtual world," all their assets and achievements will be cleared.

The emergence and maturity of the blockchain will perfectly solve this problem, enabling the metaverse to complete the evolution of the underlying architecture – blockchain can create a fully functioning economic system in the metaverse that links the real world, connecting player assets in the process. As the blockchain is completely decentralized, players can continue to invest resources.

Fortnite creator and "father of the virtual engine," Tim Sweeney pointed out that blockchain and NFTs are the "most reasonable path" to the emerging metaverse (virtual world). The NFT is the only cryptocurrency token representing digital assets under the blockchain framework; it is the economic cornerstone of the future metaverse. NFTs can be bought and sold in the same way as physical assets, ensuring the authenticity right of the underlying assets in the metaverse.

In recent years, the NFT market has grown exponentially year by year. In 2019–2020, the total global transaction volume rose from $62.86 million to 250 million – nearly a threefold increase. And 2021 was the year of the NFT. From January to August 2021, the transaction volume exploded, and OpenSea made use of its advantages in NFT users and NFT asset types to quickly dominate the market share of NFT exchanges. In August 2021, its NFT transaction volume exceeded one billion dollars, accounting for 98.3% of the global total. Comparatively, its full-year transaction volume was lower than $20 million in 2020.

From the perspective of market space, measured by NFT alone, it has reached the level of 10 billion yuan. In 2021 Q1, the NFT transaction space reached 2 billion dollars (around 12 billion yuan). After annualized processing, the annual transaction volume is estimated to be 48 billion yuan (for order of magnitude reference only), and if the metaverse scene continues to contribute 27% (refer to 2020) of its share, the space of the metaverse-blockchain link will go beyond at least 13 billion yuan in 2021.

For example, in *The Sandbox*, a sandbox game built on the blockchain, players can build their own worlds, make their own items, and develop their own games for others to use. *The Sandbox* created the SAND token, and by owning the SAND token, players can participate in the governance of the Decentralized Autonomous Organization

(DAO). And *The Sandbox* players can create NFT assets and upload them to the market. Meanwhile, Sandbox cooperates with Atari, Crypto Kitties, Shaun the Sheep and other gaming companies/IPs to create a "Play to Earn" gaming platform.

At present, users can independently manage the community in a blockchain model in *The Sandbox*. DAO is based on the core idea of blockchain. It is a form of organization derived from the collaborative behaviors of co-creation, co-construction, co-governance, and sharing spontaneously exhibited by groups that have reached the same consensus. Also, it is a subsidiary product after blockchain solves the trust problem. A DAO encodes an organization's management and operational rules on the blockchain in the form of smart contracts, thereby enabling it to operate autonomously without centralized control or third-party intervention.

A DAO is characterized by total openness, autonomous interaction, decentralized control, complexity, and diversity; it can become an effective organization to deal with uncertain, diverse, and complex environments. Different from traditional organizations, a DAO is not limited by the space of the physical world, and its evolution is driven by events or goals. It can be quickly formed, spread, and is highly interactive, and automatically disbanded with the disappearance of the target. It can help govern all business models based on blockchain and quantify the workload of each participant involved, including cryptocurrency wallets, APPs, and public chains. The main source of DAO's revenue is the collection of transaction service fees, which are usually paid through digital currency.

On the basis of DAO, Sandbox launched SAND tokens and encouraged users to create NFTs to sustain the in-game economy. The advantage of introducing digital currency and NFTs is that, unlike traditional game assets, digital currency and NFTs cannot be copied endlessly. This gives users a stronger real experience and sustains the in-game economic system without severe "inflation." On the other hand, *The Sandbox* can keep players active through NFT incentivized transactions, and fully energize their creativity.

In 2021, the application data of metaverse-NFT exploded, and it is expected to further grow in the future. *The Sandbox's* second round of land auctions in 2021 (over $2.8 million) surpassed the combined revenue of its land auctions in 2019 and 2020; its most recent land auction broke all records, reaching nearly $7 million. *The Sandbox* has a total of 166,464 pieces of land, nearly 50% of which have been auctioned. Animoca Brands, the parent company of *The Sandbox*, received an additional $88.88 million in equity financing on May 13, 2021 – on top of announcing its valuation at one billion dollars.

Consolidate the Underlying Infrastructure

The layers of spatial computing, man-machine interaction, and infrastructure are technical guarantees of the metaverse.

On the one hand, core technologies at the software level, such as 5G, cloud computing, and AI, will become key in driving the improvement in data volume/precision and accelerating the arrival of the metaverse. The China Academy of Information and Communications Technology reports that there are currently two technical paths for VR: stand-alone intelligence and networked cloud control. The former mainly focuses on fields such as near-eye display and perceptual interaction, while the latter targets streaming services after content is uploaded to the cloud. It is foreseeable that in the future metaverse framework, the two will be organically integrated on the basis of 5G infrastructure, and AI+ & cloudification together will trigger an industrial leap.

The three major branches of AI – computer vision, intelligent speech, and machine learning are playing an important role in the metaverse prototype, and Chinese manufacturers at all levels are blooming. Take the open platforms as an example: iFlytek has earned 1.76 million customers and supported 2.9 billion terminals in total; Tencent's AI open platform has 2 million customers and serves more than 1.2 billion global users. Taking development frameworks as an example, Chinese AI development frameworks such as PaddlePaddle of Baidu, MeEngine of Megvii, MindSpore of Huawei, and Jittor of Tsinghua University have achieved industrial breakthroughs and have wide coverage.

In addition, cloud rendering is also an important technology that supports the implementation of the metaverse. The development path represented by Cloud VR is divided into three stages: near-term cloudification, mid-term cloudification, and long-term cloudification. The Huawei Cloud VR service introduces cloud computing and cloud rendering technology into VR business applications. With the help of Huawei Cloud's high-speed and stable network, the cloud's visual and audio output is encoded, compressed, and transmitted to the user's terminal device. The Cloud VR development kit is mainly used for offline development; Huawei's Cloud VR connection service is adapted to the cloud of the operator's network, so that it not only can directly provide commercial services for industry users, but can be re-developed and integrated also.

At present, the AR/VR equipment industry is entering the fast lane of industrial development, and the concept of the metaverse will further accelerate equipment penetration and user cultivation. From the perspective of the equipment industry chain, the core hardware technology involves sensors, display screens, processors, optical devices, etc.; software technology mainly involves modeling, rendering, panoramic, and simulation that are content-production-related; interactive technologies have extended from traditional gamepad gestures to diverse interactive technologies such as voice, facial expressions, and eye tracking. As the structure of VR equipment gradually matures, its hardware technology will go wireless, its software technology will become cloud-based, and its interaction aims to develop into a full-scene application.

The mature technology will drive rapid growth in the VR consumer market. In the long run, driven by 5G communication, there will be richer forms of VR products. With the gradual improvement of the VR industry chain, the industry will cast a strong flywheel effect: at present, VR has been widely used in real estate transactions, retail, furnishing, cultural tourism, security, education, and medical care; In the future, VR will be fully integrated with industry.

From the perspective of the terminal equipment market, VR leads and AR follows – the 100 billion-level industrial space is being released. China Academy of Information and Communications Technology predicts that global virtual device shipments will reach 75 million units in 2024, of which 33 million units are VR devices, and 42 million are AR devices. According to Trendforce statistics, the compound annual growth rate of AR/VR shipments will be 39% over the next five years.

As the present leading brand in global VR, Facebook started to lay out VR business when it acquired Oculus in 2014. In September 2020, Oculus released a new product – Quest 2. The head-mounted device (HMD) weighs only 503 grams. Equipped with a Snapdragon VR dedicated chip, XR2, it has a refresh rate of up to 90Hz, a resolution that is 50% higher than the previous model, and larger memory and faster response speed. It was sold at only $299; its cost performance was significantly improved. According to IDC data, with the release of Quest 2, Oculus occupied 63% of the global VR market in 2020, with 3.47 million units sold. This was much higher than the debut annual sales of Oculus Rift and Quest 1 (390,000 units in 2016 and 1.7 million units in 2019, respectively). And Oculus' global shipment market share in a single quarter after releasing Quest 2 in 2020 Q4 was 82 percent.

Since the release of Quest 2, Oculus has increased the product's refresh rate from 90Hz to 120Hz through a software upgrade in April 2021. At the same time, Mark Zuckerberg, CEO of Facebook, mentioned in an exclusive interview with *The Information* on March 8 that Facebook had already begun to develop new iterative versions of the Oculus Quest, which is likely to perform eye tracking and face tracking. In addition, he also disclosed that Facebook would continue the low product price policy strategy, set up a business model around social experience, and keep expanding the user base by lowering the prices; thereby, creating more social opportunities for users to join the virtual world.

In addition to Facebook, according to *Fast Technology News*, Sony announced on February 23, 2021, that a new generation of VR system would be installed on the PS5 to optimize the user experience. From 2016 to 2020, Sony sold a total of 5.46 million

VR units (IDC data). Based on a good user base and strong demand for replacements, the second-generation PSVR is expected to become another hot VR product after Quest 2.

Meanwhile, Chinese terminals are also exhibiting a fast popularization trend. For example, Huawei VR Glass uses an ultra-short-focus optical system with 0-700-degree diopter calibration – it is a relatively light VR device, which supports VR mobile phones and dual application screen mirroring. And it is expected to create more IoT modes with its Harmony OS. The official price of this VR Glass in 2019 was 2999 yuan. And one major selling point of it is the content side. There were rich scenes, including panoramic video, virtual IMAX, regular film and television channels, and over 100 selected VR games were released simultaneously.

6.3   The Plan about the Metaverse

Internet giants, with their own enormous traffic portals, race to seize various layers of the metaverse industry chain, so as to hold the metaverse ecosystem platform to directly connect with users.

Facebook: In-depth Arrangement of VR, Leading the Social Metaverse

Facebook is the world's leading provider of online social media and web services. It was launched on February 4, 2004. In order to create a multi-category and differentiated social matrix, the company has made several investments, mergers, and acquisitions. In 2020, it achieved growth in both revenue and profit. Through Oculus devices, it set foot in the VR field, and launched *Facebook Horizon* as a VR social platform. At present, Facebook has become the most well-known metaverse-concept company in addition to Roblox. Facebook also changed its name to Meta in October 2021.

In 2014, Facebook acquired the VR company Oculus for two billion dollars, joining the VR industry. In its ten-year plan in 2016, Zuckerberg declared that in 3-5 years, it would focus on building a social ecosystem and completing the functional optimization of core products; for the next 10 years, it would be committed to new technologies such as VR, AR, AI, and drone networks. By 2021, the proportion of Facebook employees involved in VR/AR technology R&D has increased from 10% in 2017 to 20%, and it has frequently invested in technology leaders in the VR/AR field.

At last, its long-term plan for VR hardware terminals has paid off. The demand for Oculus Quest 2 (now known as Meta Quest 2) is strong. Quest 2, thinner and lighter than the Quest, has four tracking cameras and two black-and-white Oculus touch motion controllers on its front. It doesn't allow users to keep completely separate Oculus accounts, and Quest 2 product managers explained that this was to improve the sociality that Quest offers. Users can easily find friends in VR by linking their Facebook accounts, and chat virtually with them using Facebook Messenger from their Quest devices.

Moreover, Facebook is redefining office space by creating the VR office "Workrooms" on Quest 2. Users can participate in virtual meetings in the form of avatars. Such virtual face-to-face communication can greatly improve remote meetings as well as the efficiency of brainstorming and some other creative scenarios.

Workrooms provides a mixed VR experience. There, users can express their ideas on various virtual whiteboards, and bring their desks, computers, and keyboards into the VR world for normal office work. Oculus Avatar offers users richer appearance choices too, and Workrooms provides various office scenes and furnishings.

According to IDC data, in the global VR HMD market in 2020 Q2, Facebook replaced Sony as, the company with the largest market share – 38.7%. Meanwhile, the demand for the newly released Oculus Quest 2 in 2020 has been strong: it was soon out of stock in the North American market. And the annual shipment of this device is expected to be 5–7 million units. According to SuperData statistics, it dominated the VR market in the fourth quarter of 2020, with sales of about 1.1 million units – far ahead of the second-placed Sony PSVR.

At present, the beta version of *Horizon*, a VR social platform on Facebook, is applicable to gaming, social networking, and efficient office work. It is therefore called the Roblox of the VR field and is believed to be an important step for reaching the metaverse. *Horizon* allows up to 8 players to create their own virtual worlds together on the platform. Players create and decorate the "worlds" with their own virtual cartoon avatars, and play various social mini-games.

The head of product marketing at Facebook Reality Labs Experiences believes that one of the important components of the metaverse is a platform with more opportunities for social interaction in a VR environment – one that makes the social engagement in VR deeper and wider. With the dual advantage of the technical field and the social field, Facebook is expected to build an enormous social metaverse platform.

In addition to endeavors in VR and the development of a VR social platform, Facebook's plan for the metaverse also includes Creator (a content creation community), Spark AR, and diem (a digital currency).

Creator is an app designed to allow content creators to build a community around content and provide one-stop creation services, including video creating, editing, and publishing; receiving information and comments from apps such as Instagram and Messenger through Creator; sharing information on Facebook through Creator; sending content to Twitter, Instagram and other platforms to help creators perform statistical analysis and publish more popular videos.

Facebook also released Spark AR for Instagram, describing it as a platform where anyone can create and publish AR effects. The published filter effects will be included in the Instagram library of new effects. Recently, Spark AR has added two new functions: multi-level segmentation and optimized target tracking, which enhance the recognition level and target number of AR for better realistic effects.

The virtual digital currency diem is similar to Tether and other price-linked stablecoins (diem is linked to the US dollar). Supported by traditional assets, it runs on the diem project's own blockchain, and is stored in the Novi wallet. The diem blockchain can be made, and like Ethereum developers can create customized applications. The market value and circulating supply of diem are not fixed; the diem association can make or destroy diem as dollars move in and out of its collateral reserve. And the metaverse properties of diem depend on its pay ability in the real world. It is likely that some members of the diem association will accept it as a payment method, such as Shopify, Spotify, and Uber.

Tencent: Internal Incubation and External Investment – A Complete Internet Reality

Tencent is a Chinese social networking and online entertainment leader, and its business scope covers the entire Internet ecosystem. Tencent is one of the largest Internet integrated service providers in China. It owns WeChat – a social platform with one billion daily active users, and *Honor of Kings* – a mobile game with over 100 million daily active users and the world's highest game revenue.

Tencent's principal business is divided into three major parts according to the composition of income: first, value-added services, which consist of two parts: online games and social life. In terms of online games, self-developed games *Honor of Kings* and *Game for Peace* have long topped the chart of best-selling mobile games. In terms of social life, it renders membership services for music and videos, online literature, and live streaming. The second is online advertising, which mainly includes media advertising based on Tencent Video and social advertising based on platforms such as WeChat and QQ. The third is financial technology, which mainly includes Internet finance, cloud and enterprise services, etc.

Tencent is taking games as its entry point: through internal incubation and external investment, it is drawing up plans for multiple metaverse fields. In fact, Ma Huateng had already described it in Tencent's internal publication, *World, Values, Life*, as: "An exciting opportunity is coming, the mobile Internet that had been developing for ten years is about to welcome the next wave of upgrades, which we call the complete reality of Internet. This is a process from quantitative change to qualitative change, which means the integration of online and offline, the sub-merge of physical and electronic means. The portal between the virtual world and real world is open. Whether it is from virtual to real, or from real to virtual, both are dedicated to giving users a more realistic experience." This is a highly accurate description.

In terms of internal incubation, Tencent is actively exploring the in-depth integration of games and social interaction, and deploying an entire industry chain in the metaverse. In terms of the upstream content ecosystem and self-developed games, it continues to develop open-world games, including the production sandbox game *Our Planet*, and the open-world mobile game *Awakening of Dawn* produced by UE4. Their high degree of freedom is similar to the open source and creation characteristics of the metaverse.

Additionally, on April 15, 2021, Tencent Platform and Content Group (PCG) announced the appointment of Yao Xiaoguang, president of Tencent Interactive Entertainment TiMi Studio, to take over the overall business of the PCG social platform – whose two major products are QQ and QZone. In January 2020, Tencent COO, Ren Yuxin, stated in an internal letter that PCG shoulders the heavy responsibility of Tencent to explore the future development of digital content.

Yao Xiaoguang was one of the earliest senior programmers engaged in online game R&D in China. He founded the website npc6.com, supervised the development of *Monster & Me* – the first turn-based MMORPG game in China, once worked as the executive deputy general manager of Shanda Network Shengjin Branch, and participated in the core development of many large games, such as *The World of Legend*. Meanwhile, he has written and published many professional books, providing a lot of theoretical guidance for the training of strategic and technical talents in China. Mr. Yao joined Tencent in 2006 after rounds of sincere invitations from the company.

In the early days of Tencent, he was responsible for the R&D and operation of Gem in the Sky Studio. QQ Speed, the first project that he led, made him and the studio famous. Prior to this, Tencent had zero experience in large self-developed games, and mainly relied on agent service. In October 2014, the Gem in the Sky Studio, TiMi Studio, and Wolong Studio were merged into the TiMi Studio Group. And Yao's representative works of PC-end games are *QQ Speed*, *X-Game*, *Assault Fire*, etc.; representative works of mobile-end games *are Cool Running Every Day*, *Fighting Days*, *Crossfire: King of Shooting*, *Honor of the Kings*, etc.

In the face of fiercer competition in Internet content and channels, Yao shoulders the burden of saving QQ – combining games and social networking, and creating Tencent's new trump card. Obviously, through the deep integration of games and social interaction, Tencent will further explore the product ecosystem of the metaverse.

In terms of downstream infrastructure, Tencent plans to invest $70 billion in cloud, AI, blockchain, 5G, and quantum computing in the next five years. By improving game accessibility, reachability, and scalability, it approaches the mature form of the metaverse.

Tencent cloud technology and financial technology are the underlying support of metaverse development. With the accumulation of user traffic and products on social platforms, payments, and other fields, Tencent cloud computing holds great advantages in the consumer Internet and finance. In the future, the potential of Tencent Cloud will be mainly concentrated in the SaaS field. Also, it will serve as an important strategic layout.

Tencent has made some moves in blockchain, too. On August 10, 2020, Tencent Music began reserving its first batch of digital collections (the 20th Anniversary Vinyl NFT of Hu Yanbing's single, *Monk*). Users could open a reservation lottery for purchase qualifications on QQ Music, and a limited number of 2,001 copies were sold. The lottery time was 10:00 am on August 14th, and the official release time was 10:10 am on August 15th. Tencent Music became the first music platform in China to distribute digital collection NFTs.

As for financial technology, compared to Alipay, WeChat Pay is ahead in terms of daily average payments by virtue of its own traffic base and ecosystem construction.

As Tencent's financial business' product line continues to improve, it is expected to steadily narrow the gap and catch up with Alipay. High-profit businesses such as wealth management and small loans will contribute more income under the gradual establishment of the credit system. One of the focuses of the metaverse concept is digital economic civilization. And Tencent's financial technology business has the potential to build a virtual currency system.

In terms of external investment, Tencent has invested heavily in infrastructure to complete the metaverse picture. By 2020, it had invested in over 800 companies, many of which are metaverse-related. In the gaming arena, it continues to invest in companies and products related to the metaverse concept: Roblox and Epic Games are both on its list. Tencent has invested in Epic Games since 2012, whose Unreal Engine is widely used in the development of simulated games. In terms of VR/AR components, Tencent has invested in Snap since 2013, which leads the industry in the manufacture and application of AR components.

In the digital economy, the game virtual currency Robux, introduced by Roblox, one of Tencent's investees, can be exchanged for cash through the game platform, thus connecting real and virtual currency. In addition, Tencent has also invested in a number of e-commerce companies at home (Pinduoduo and Meituan, etc.) and abroad (Paystack and SeaMoney, etc.), whose social shopping model is expected to help the construction of a virtual closed-loop economic system. On the whole, Tencent's diverse internal incubation and external investment layout make it one of the most likely companies to build a metaverse prototype.

ByteDance: Acquire VR Leaders to Enter the Metaverse

In 2014, under Zuckerberg's judgment that future consumption of VR is comparable to that of mobile phones or PCs, Facebook paid two billion dollars to acquire Oculus and entered the VR arena. Thereafter, Oculus has grown rapidly. It was estimated that

Oculus Quest 2 would sell 8-9 million units in 2021. Seven years later, ByteDance, which is comparable in size to Facebook, acquired Pico – a Chinese brand comparable to Oculus.

ByteDance's acquisition of Pico is seen as a remake of Facebook's acquisition of Oculus. Although it took place 7 years later, in terms of the market, ByteDance's entry was just at the right time. In the past 7 years, the VR industry has suffered two giant bubbles. It is not until now that the technology involved and market demand have matured.

Founded in April 2015, Pico is a VR software and hardware R&D manufacturer.

In December 2015, Pico launched Pico 1 VR HMD, Pico VR APP, and Pico industry solutions.

In April 2016, the VR all-in-one machine – Pico Neo DK – came out.

In May 2017, it released the display dock product, Pico U, an upgraded version of the split-type VR all-in-one Pico Neo DKS, the flagship all-in-one product Pico Goblin, and the VR Pico Tracking Kit.

In December 2017, the first mass-produced 6DoF Pico Neo VR all-in-one was released.

In July 2018, the Pico G2 VR all-in-one, using the Snapdragon 835 chip and priced at over 2000 yuan was released; Pico received 167.5 million yuan in Series A financing.

In May 2019, the upgraded Pico G2 4K came out.

In March 2020, the second-generation high-end 6DoF all-in-one, Pico Neo2, was launched.

In May of that year, the new generation Pico Neo3 came out.

So far, Pico has developed products such as the Goblin VR all-in-one, Pico U VR glasses, and Tracking Kit. This company of 300 employees has branches in Tokyo, San Francisco, and Barcelona as well as an office in Hong Kong. And its offline sales channels cover over 40 Chinese cities in seven regions.

It is worth mentioning that Pico is one of the few companies to have survived the capital winter in the last VR wave. Its founder Zhou Hongwei was a Goertek executive.

(Goertek is both a shareholder and a supplier of Pico, and the optics and hardware of all Pico products are supplied by Goertek.) It is also one of the main foundries for the Oculus Quest series.

China International Capital Corporation believes that when Pico is merged into ByteDance, it is expected to integrate ByteDance's content resources and technical capabilities, and continue to increase investment in VR product R&D and the ecosystem. From the perspective of shipments, current VR development is still dominated by Western markets. Facebook has enriched the game content ecosystem and enhanced user stickiness by acquiring Oculus and VR game teams. ByteDance's acquisition of Pico is expected to become a turning point for the development of the Chinese VR market. Relying on Pico's relatively mature hardware ecosystem, together with ByteDance's software development capabilities, the prosperity of Chinese VR is promising.

It is worth mentioning that Pico, which has attracted wide attention from the outside world, is not the first project that ByteDance has planned in the metaverse field. As early as April 2021, it invested 100 million yuan in Beijing Code Qiankun Technology, a game developer known as the Chinese Roblox.

Founded in 2018, Beijing Code Qiankun Technology is a game platform with UGC, and its representative work is the metaverse game *Reworld*. Based on its self-developed interactive physics engine technology system, the company launched the UGC creation platform *Reworld*, which mainly consists of two parts: the physics engine editor (PC) and the game work sharing community (App). It enables users to freely create models, publish their own gameplay, model materials, and finish games or store them for other developers or players to use. At present, *Reworld* is the only one UGC game creation platform with self-developed physics engines in China, in addition to the imported Roblox.

6.4 Tech Giants Scramble to Corner the Metaverse Market

Needless to say, building the metaverse is a massive mission. It requires a high-speed, low-latency, a super-connected communication environment, massive data processing, cloud real-time rendering, smart computing, etc. However, with recent technological progress in 5G, big data, and AI, this ethereal concept of the past is now possible.

Huawei: Master of the Underlying ICT Technology

Huawei Cloud, as the fastest growing cloud, stood among the top five in the world last year. It mainly covers the public field of cloud computing, providing basic cloud services such as cloud hosts, cloud trust, and cloud storage, as well as solutions such as super-computing, content distribution and acceleration, video hosting and publishing, and cloud conferencing. Gartner reported in April that its ranking in the global cloud computing IaaS market in 2020 rose to the top two in China and the top five in the world. With a growth rate of 168%, it registered the fastest growth among mainstream service providers.

According to IDC data, in the first half of 2020, Huawei Cloud's Model Arts held the No.1 market share (29%) of machine learning public cloud services in China; among industrial cloud solution providers, Huawei Cloud also ranked second with a share of 11.5%; in the container software market, it came out top in Chinese market share.

Huawei is also actively exploring the VR field, and continues to promote the construction of the AR/VR ecosystem. It provides a platform specifically for VR content developers – HUAWEI VR. Developers can use Huawei VR SDK to create. After the work is completed, they can upload it to the Huawei VR app store. And consumers with Huawei VR glasses can directly download it. The current hardware form of Huawei VR is VR glasses + Huawei mobile phones/tablet PCs. VR phones come with 3DOF HMD and 3DOF controllers.

Huawei's technological breakthroughs in the AR/VR field have accelerated the realization of immersive experiences. It launched the *Cyberverse* underlying technology platform, which includes full-scene spatial computing capabilities, AR walking navigation, scene editing and rendering, etc. At present, *Cyberverse* has been applied to the panoramic resurrection of the Mogao Caves in Dunhuang, achieving a wondrous combination of technology and culture. In the meantime, Huawei also released the general-purpose Huawei AR Engine, which allows developers and third-party applications to access Huawei's VR system.

In the future, with the further advancement and fusion of 5G and cloud computing, this model is expected to lead VR industry development, or even become a key step for mankind to enter the metaverse.

Nvidia: Release of the Groundbreaking Omniverse Platform

Omniverse is an open cloud platform developed by Nvidia for virtual collaboration and real-time full-fidelity simulation. Through the cloud, creators, designers, engineers, and artists can work together locally or anywhere beyond physical boundaries, and see their results in real time. This is far more convenient. As a cloud platform, Omniverse has a high-fidelity physics simulation engine and high-performance rendering capabilities. It supports multi-persons to co-create content on the platform, and is highly compatible with the real world. With proper data, it enables the creation of a virtual world exactly the same as the real one.

Through the 3D exchange capability, ecosystem members are connected to large user networks, 12 connectors are available for mainstream design tools, and an additional 40 connectors are in development. Its early experience partners come from a variety of industries around the world, including media & entertainment, gaming, AEC (architecture, engineering, construction), manufacturing, telecommunications, infrastructure, and automotives.

Obviously, Omniverse's vision agrees with one of the most important metaverse concepts: it is not operated by a single company or platform, but works in a multi-party, decentralized way.

Goertek: The Number Founder of AR/VR Equipment

Goertek, widely known as the leader of the Apple industry chain, has long been involved in the metaverse concept. Goertek, founded in 2001, mainly engaged in the R&D, manufacturing, and sales of acoustic-optical precision components and structural components, intelligent machines, and high-end equipment.

In the past three years, its revenue and net profit have continued to increase by a greater amount each year. Though it performed poorly in 2018, with negative growth in revenue and net profit, it recovered fast. In 2020, its annual revenue and net profit grew by 64.29% and 122.41% respectively. In addition, its R&D expenditure has climbed steadily in the past five years, but its gross profit margin has suffered a slight drop. From 2016 to 2019, the year-on-year growth rate of R&D expenditure exhibited a downward trend, but it surged to 74.64% in 2020. That year, its gross profit margin exhibited a downward trend as a whole, and the gross profit margin was 16.03%.

In the AR/VR field, Goertek's plan is mainly reflected in parts supply and complete machine assembly. Precision components are mainly combined with the key optical devices of VR/AR for development. Goertek occupies 80% of the global market share of mid-to-high-end VR products.

In 2020, its smart hardware revenue, including AR/VR business, accounted for 30% of its sales revenue. In 2020, its performance grew against the market trend, with its AR/VR business making an outstanding contribution. However, the biggest revenue contribution came from the smart optical complete machine business, accounting for 46.20%.

At the end of 2020, when the market expected Apple's wireless earphone sales to decline in 2021, the 21st Century Capital Research Institute pointed out in a report that the VR/AR business may take up the mantle from headsets, becoming a new profit source for Goertek. Since 2021, Goertek's business performance has proved this to be true.

According to goer's 2021 annual report, in 2021, the company's annual operating revenue was 78.221 billion yuan, a year-on-year increase of 35.47%, and the net profit was 4.275 billion yuan, a year-on-year increase of 50.09%. On April 26, 2022, goer released the first quarterly report of 2022, which showed that the company realized operating revenue of 20.112 billion yuan in the quarter, a year-on-year increase of 43.37%; The net profit attributable to the parent company after deduction was 878 million yuan, with a year-on-year increase of 46.06%.

Goertek pointed out in its 2021 mid-year financial statement that emerging concepts like the metaverse are attracting wider attention across the industry. In the first half of the year, it performed far better thanks to the increase in sales revenue, including VR equipment, smart wireless earphones, etc. In 2021, the revenue of intelligent hardware was 32.8 billion yuan, accounting for 41.94%, an increase of 85.87% over the same period in 2020. In 2021, the two hard currencies of PS5 and Quest 2 sold well all over the world. Goertek is also the OEM of meta Quest 2, PS5, and Pico products. Its smart hardware business led by VR/AR demonstrated eye-catching and explosive growth.

CHAPTER 7

The Metaverse Needs a Charter

The ultimate form of the metaverse will be the perfect fusion of openness and closedness. Like our universe, it is where open and closed systems coexist and are even partially connected, where big and small universes are intertwined. Ultimately, multiple metaverses of different styles and fields will make up a larger metaverse, identities and assets of users will be synchronized across the metaverses, and lifestyles, production modes, and organizational governance patterns will be reconstructed. This full-scale metaverse will take human civilization into a brand-new digital age, but before this ideal form arrives, people still need to create a lasting charter for it.

7.1 Towards the Future Metaverse

At present, metaverse-related topics are attracting great market attention. There are both the voices of disagreement and consensus. As a basis and prerequisite, hardware equipment like VR continues to improve its experience but lower its prices. Under the promotion of major manufacturers, their market penetration is expected to welcome a surge.

5G, cloud computing, blockchain, and other infrastructure are gradually maturing, and under the impetus of hardware and infrastructure, content and application ecosystems are also developing rapidly. In the future, our understanding of the metaverse will deepen, and from the consumer Internet to the industrial Internet, the metaverse era of online and offline integration will be gladly accepted.

Short Term: Technology Catalyzes Development

The evolution of cutting-edge technology has made the metaverse possible, and the rapid development of the Internet has built a solid foundation for it, which is essentially a virtual network world. When the hardware foundation is laid, it requires as many users as possible to constitute the behavioral agents and content creators in the metaverse.

Therefore, in the short term, the metaverse has to go through continuous technological advancement. Indicators for the completion of this stage include but are not limited to: 5G market penetration reaching 80%; cloud gaming and XR achieving mature application; top-level gaming companies represented by Tencent making breakthroughs in next-generation games; and AI enabling AI-assisted content production. In major economies such as the US and China, there are a number of immersive experience platforms that combine games, social networking, and content – their market penetration exceeds 30%.

The metaverse concept will remain focused on the recreational field, such as games, social networking, and content. With the upgrade in communication, computing power, VR/AR devices, and AI, experiences will become more immersive. And this experience is one of the most important forms at this stage.

In terms of software, the metaverse unfolds based on the UGC platform ecosystem and social platforms that can build a virtual relationship network. The underlying hardware support cannot do without the mobile devices widely in use today. Meanwhile, major Internet giants and some leading companies focused on games and social networking will develop a series of independent virtual platforms, which are expected to become carriers of new recreational lives.

In the social realm, the metaverse will continue to provide users with a social experience that combines gameplay and avatars. In fact, the core social experience currently provided by the metaverse lies in the highly immersive social experience and rich online social scenes brought about by gameplay.

A virtual identity tears down social barriers and give users a stronger sense of substitution; through personalized virtual identities, users can customize their look. For example, Roblox runs a rich Avatar store. At the same time, virtual social platforms remove a range of social barriers, including factors such as physical distance, appearance, wealth gaps, or racial and religious differences, so that users can freely express themselves.

Meanwhile, various user game behaviors actually carry social functions. For example, in *World of Warcraft*, players carry social attributes, and social interaction takes place through battlefields, dungeons, and other modes. Furthermore, the team dungeons and faction battles in games such as *World of Warcraft* and *JX Online 3*, the multi-player team formation in competitive games such as *Honor of Kings* and *Game for Peace*, and the release of games like *Mole's World* encourage players to engage in social activities. In addition to game interaction, *Roblox* and *Fortnite* both have party modes so that players can host parties or concerts in the virtual world, and *Mole's World* collaborated with the Strawberry Music Festival and invited the band New Pants to party online.

Social networking is clearly an important means of demolishing the boundary between the virtual world and the real world. As the underlying technology advances and the social scenes broaden, the immersion and fidelity of simulation brought about by the metaverse will be further upgraded. For example, the social software, *Soul*, builds a virtual world for users. Users socialize through virtual identities on the *Soul* platform. As social barriers are gone, users have more freedom to express themselves. Simultaneously, Soul users can have group chats, listen to music, study, play games like *Werewolf*, and even purchase real goods for themselves or others through Giftmoji.

In the content field, the metaverse hopes to create an extremely realistic virtual universe. By nature, it continues to expand, from orderly to disorderly, and fundamentally demands increased content volume and production. Only when the content reaches a large enough volume can it be called the metaverse. Therefore, in the short term, it will keep expanding.

At present, many film companies, comics, and other content producers are trying to create their own IP universes by building a "worldview." For example, the most successful Marvel Universe production for now – *Iron Man* in 2008 – has gone through three stages, while the movie *Black Widow* opened the fourth stage of the Marvel Universe. In the past 13 years, 23 movies and 12 TV series have been produced. Marvel films are made on the alternate world of Marvel Comics, and belong to the same officially recognized multiverse. From comics to single-hero movies, to the interactive development of multiple heroes, Marvel has also strengthened the market penetration of its cosmic ecosystem from various derivatives, such as games and offline theme parks. A single IP or multiple independent IPs cannot constitute a universe, but a series of IPs and strong correlations between them (that is enriched through various forms of content, and a series of secondary creations by users) can.

Based on this is Tencent's idea of pan-entertainment – the all-round supply and continuous content derivation of the industry chain. Tencent may rely on its strong social networking influence to actively plan in the pan-entertainment sector through internal incubation and external investment, and become a leader in online literature, animation, online music, film and television production, video platforms, and online games. Gradually, it is going to build an entertainment matrix with wide coverage and huge influence around IP.

The boundary of the metaverse keeps expanding. A16Z divides the evolution of content production into 4 stages. At present, we have entered the stage of UGC from that of PGC. Both content production capacity and mainstream social forms have achieved leapfrog improvement. For example, in the open-world game *GTA*, the boundary of pure first-party game content is still limited by the production capacity of professional teams. However, with the emergence of MOD (modification) made by players themselves, game content can be added or replaced, thereby greatly enriching its content.

UGC is the first-level detonator of the content ecosystem. For instance, head content platforms Douyin, Kuaishou, and Bilibili, in addition to some professional PGC producers, have formed an expanding content library – the content production capacity of some UGC has even reached that of PUGC. Among them, the number of active content creators on Bilibili in 2021Q1 was 2.2 million, with a year-on-year increase of 22%; while the number of monthly average high-quality video uploads reached 7.7 million, with a year-on-year increase of 56%; the number of average monthly video uploads for a single active Bilibili-er increased to 3.5 – a month-on-month increase of 0.5. This helped drive Bilibili's average daily video view counts to reach 1.6 billion, with a year-on-year increase of 45%.

Additionally, the production of lots of high-quality UGC also requires the introduction of AI-enabled content creation. At present, companies are exploring this. For instance, Roblox is using machine learning to automatically translate games developed in English into eight other languages, including Chinese, French and German. Meanwhile, companies such as Xinhua News Agency in collaboration with Sogou, ByteDance, Baidu, iFlytek, and others, have launched interactive AI anchors. While we are still in the development phase of AI, the upgrade and adoption of AI tools can make the content creation easier, thus allowing producers to focus on content quality. With the continuous penetration of AI, future content production is expected to eventually become full AI.

As technical merits improve, the immersive experience of future content is expected to get better. Compared with traditional videos, content in the metaverse era will be presented in a more realistic and in-depth fashion.

In terms of film and television, they will be presented in the form of AR/VR interaction; combined with the multi-person social interaction mode, immersive online live-action role play will be created; through AI, a true open plot and multiple sub-plots are created, and the matching plot will be selected according to the player's preference, etc.

In terms of music, there will be immersive music videos combined with a Karaoke mode, and people will have the direct opportunity to perform on the virtual stage with their favorite singers and idols.

Long-term: Penetrate Production and Life

The long-term metaverse trend is actually an open question. Though various cutting-edge technologies are advancing fast, and human demand is picking up, there are still many uncertainties. Like us back at the end of the 20th century – no one would imagine that 30 years later, there would be a smartphone, paperless offices, open social networking, and digital shopping.

Regardless of so many uncertainties, there are still traces of the metaverse development path. Clearly, penetration will mostly take place in areas that can improve production and life efficiency. VR/AR and other display and cloud technologies, smart cities, a virtual consumption system that becomes a closed loop, virtualized service forms, and a financial ecosystem of matured digital assets will constitute the important parts of the metaverse. And the development of blockchain has bridged between the top and bottom layers of the metaverse.

Clearly, a complete and robust technical system is needed to support its governance and incentives: based on its own technical characteristics, blockchain naturally adapts to the key application scenarios of the metaverse. It can be applied to digital assets, content platforms, game platforms, the sharing economy, and social platforms.

On the one hand, a key feature of metaverse governance is its joint construction by countless centralized institutions and individuals, so it should be distributed, decentralized, and self-organizing. On the other hand, this feature ensures that digital assets cannot be copied, thus the metaverse community can remain stable. With blockchain, the metaverse participants can receive rewards based on their contributions (time, money, content creation) in the metaverse. And NFT, a token that can be used to represent the ownership of unique items, will act as a medium for this.

The digital scarcity that comes with NFTs is ideal for collectibles or assets whose value depends on limited supply. As aforementioned, Crypto Kitties and Crypto Punks are two of the earliest NFT use cases, with a Crypto Punk NFT – Covid Alien – selling for $11.75 million. In 2021, popular brands like NBA TopShot were trying to create NFT-based collectibles that contain video highlights from NBA matches rather than still images.

NFT enables artists to sell work in its natural form without having to print them out. Also, unlike physical art, artists can make money through secondary sales or auctions, which ensures the original work is recognized in subsequent transactions. Markets are dedicated to art-based NFTs. For example, Nifty Gateway 7 sold/auctioned over 100 million dollars of digital art in March 2021.

NFT also opens a big window of opportunity by introducing the chance of ownership. While people spend billions of dollars on digital game assets, such as avatars or outfits in *Fortnite*, consumers don't necessarily own those assets. NFT will allow players of crypto-based games to own assets, earn them in games, port them outside the games, and sell them elsewhere (like open markets).

When the economic system of the metaverse improves in the future, this will be welcomed by pan-entertainment immersive experience platforms. Consumption, education, meetings, work, and other behaviors will be at least partially transferred to the virtual world. Simultaneously, when consumption behavior in the virtual world continues to heat up, it will in turn drive some virtual platforms to realize transactions, social interaction, etc.

In the future, each virtual platform, as a sub-universe, will gradually form a complete set of standard protocols to aggregate all sub-universes and create the true metaverse. These sub-universes still maintain their independence, but are interconnected via standardized interfaces. The metaverse will enter the complete reality of the Internet stage.

7.2 Creation Dilemma

The metaverse is a beautiful vision of the future Internet. However, under current technical conditions, we have only just entered it. At present, the development of the metaverse still faces many difficulties. In order to better realize the low-latency and convenience, continuous breakthroughs in areas such as communication and computing power, interaction modes, content production, economic system, and standard protocols are required.

The Long Road to Eliminate the Digital Divide

Having benefited from the development of the digital society, the metaverse is an advanced form of the present Internet. Global Internet users have maintained high growth in the past decade. According to Zhiyan consulting, in 2021, the total global population reached 7.8 billion and the number of global Internet users reached 4.8 billion. As of January 2022, the number of global Internet users reached 4.95 billion, a year-on-year increase of 4%, and Internet users accounted for 62.5% of the total population.

The expansion of social platforms has resulted in the supporting metaverse framework. For the metaverse to ultimately realize the interaction of multiple individuals in the virtual world, social platforms must play a key role. Global social platforms are expanding fast: the Digital 2021 Report jointly released by We Are Social and Hootsuite in January 2022, shows that the current number of social media users has reached 4.2 billion, exceeding the previous year's corresponding figure by 490 million. With an increase of over 13% year-on-year, it accounted for more than 53% of the world's population.

Both the number of active users of social platforms and their average daily usage time have grown considerably. The Digital 2021 Report shows that in 2020, users aged 16–64 spent an average of 2 hours and 25 minutes per day on social media. Among global mainstream social platforms, six platforms hold over one billion monthly active users, and the number for the top 17 platforms exceeds 300 million.

Meanwhile, the development of the metaverse is also subject to the development of the digital society, which is faced with the huge obstacle of digital divide. The digital divide is a multi-dimensional and complex phenomenon that is widely present in both developed and developing countries. This concept was proposed as early as the 1990s, and with the increasing application of the Internet, it has become a general label or metaphor to describe the gap in the adoption and use of information communication technologies, especially the Internet, between certain countries or regions.

Globally, only just over half of families (55%) have an internet connection, according to UNESCO. In developed countries, 87% of the population has access to the Internet, while in developing countries, the proportion is 47%; in the least developed countries, the Internet connection rate is only 19%. According to these statistics, a total of 3.7 billion people in the world do not have Internet, and most of them are from poorer countries.

Also, in some countries, the high equipment cost keeps some people from owning a mobile phone. In sub-Saharan Africa, the cost of 1GB of data, which suffices to play an hour of standard definition movies, is close to 40% of the local average monthly salary. According to the World Bank, 85% of people in Africa live on less than $5.50 a day. Most Africans consider themselves separated by the digital divide.

In addition to the international digital divide, both developed and developing countries suffer varying degrees of domestic digital isolation. In the US, for instance, more than 6% of Americans (21 million), do not have access to the high-speed Internet connection. In Australia, the percentage is 13%, and nearly a third of low-income families are not even connected to the internet. These numbers show that, even in the world's richest countries and families, not everyone has access to Internet service.

More than 157 million Americans don't have access to the internet at a broadband speed, according to Airband, a study by Microsoft's rural Internet project. Without a proper broadband connection, these people may not be able to start or run a modern business, use telehealth, access online education, digitize farms or conduct academic research online.

In China, information technology has begun to penetrate into all industries of the national economy. The focus of digital development has shifted from consumption to production, and digitalization has become the center of industrial transformation and upgrade. In 2018, the scale of Chinese industrial digitalization exceeded 24.9 trillion yuan, accounting for 27.6% of the country's GDP. However, in the tertiary industry, the digitalization of agriculture is lagging behind, with a slow growth of digital added value. In 2018, the digital economy accounted for 35.9%, 18.3%, and 7.3% of industrial added value in China's service, industrial, and agricultural sectors, respectively, up by 3.28%, 1.09%, and 0.72%, respectively, from 2017. Agriculture lagged significantly behind.

On the other hand, the digital divide between underdeveloped and developed regions, as well as between urban and rural areas, has not disappeared with the rapid development of the Chinese economy. In April 2020, China Internet Network Information Center released the 45th Statistical Report on Internet Development in China, showing that as of March 2020, China's Internet penetration rate had reached

64.5%, while in rural areas, it was only 46.2%. The number of rural netizens was only 39.3% of the urban netizen population, accounting for 59.8% of total non-netizens.

There is no doubt that the impact of the digital divide is wide and profound. Its existence and continuous expansion will lead to an unequal distribution of benefits based on the digital economy, and prevent society from truly entering the metaverse era. Therefore, the ultimate goal of reaching every individual on a global scale must be realized in the development of the metaverse.

Technical Shackles to Break

The rapid iteration and advancement of digital technologies represented by 5G, cloud computing, AI, and VR/AR have provided technical support for the emergence of the metaverse. But in the meantime, its further development is shackled by the current technological level, and technical strength still needs to be improved.

There is a cycle to the advancement of all technologies. This theoretical framework, proposed by the Gartner Group in 1995, is being used to analyze, predict, and deduct the maturity and evolution speed of new technologies, as well as the time required to mature, so as to track their evolution; it consists of five stages, known as the Hype Cycle, showing how the cycle of new technologies is consistent and follows this five-stage pattern.

The first stage is the technology trigger, which means the birth of new technology. Often, this new technology is widely reported by major media due to its novelty and hi-tech content. The second stage is the peak of inflated expectations. Some companies launch products at this stage. Some succeed, and some fail and suspend innovation. The third stage is the trough of disillusionment, where new products and services fail to meet public expectations. Once the fourth stage – the slope of enlightenment – arrives, the new technology develops steadily, matures, and finally enters the mainstream market. This is the fifth stage – the plateau of productivity.

As Mark Raskino, author of *Mastering the Hype Cycle*, put it: "People are often excited about the possibility of a new idea coming true, because it means a possible huge impact on reality, but sometimes companies need to realize that turning an idea into reality is extremely difficult … Sometimes it takes years to solve the problem. After the third stage of the trough of disillusionment, there is not much left in the market, and those that survive are often remodeled, repackaged, or reinvented."

Both VR and AI go through such a cycle. VR was regarded as a sunrise industry in 2016, and was included in many national policy documents such as the Informatization Program in the 13th Five-Year Plan. Thus, Chinese manufacturers entered the market one after another, and the entire industry was booming. However, due to immature technology and high prices, the industry endured a cold winter in 2017–2018. It was not until 2019 and 2020 that it re-entered a period of rapid development, with the VR content ecosystem improvement and hot sales of Oculus products.

Since it was born in 1956, AI has become a leading technology for research. Even in the 1960s, when human intelligence, such as abstract thinking, self-recognition, and natural language processing, was still out of reach for machines, researchers remained optimistic about AI. Scientists at the time believed that fully intelligent machines would appear within two decades. Therefore, AI research received almost unconditional support at the time. J.C.R. Licklider, then-director of ARPA, believed that his organization should "fund people rather than projects," and allow researchers to explore any direction that interests them.

But the good times didn't last long and the first cold winter for AI came soon. In the early 1970s, AI began to suffer from criticism that even the most brilliant AI programs could only solve the simplest part of the problems they were trying to solve. This led some to conclude that all AI programs were mere "toys." AI researchers have encountered fundamental obstacles that cannot be overcome. There are also financial dilemmas that follow, and AI researchers which failed to correctly judge the difficulty of their research subject: the overoptimism before has raised public expectations too high, so when promises could not be fulfilled, the funding is curtailed or cancelled.

The present metaverse, as the integration of many technologies, is not mature. It appears to be stuck in the first stage of the Hype Cycle, receiving much attention but developing slowly. The metaverse will have to go through many trials, and its development may be harder, more expensive, and slower than expected.

Absence of Standard Protocols and Economic Systems

Standard protocols and economic systems are key elements for the metaverse to integrate countless sub-universes. Similar to the TCP/IP protocol and the TD-LTE standard for the PC Internet and mobile Internet, the formation of the metaverse requires a complete set of standard protocols, including a series of general standards and protocols for user identity, digital assets, social relations, application API, etc.

The presence of standard protocols enables user identities to be interoperable on the platforms (sub-universes) of major companies. Simultaneously, their possessions of digital assets and content need to be interoperable, too. In addition, APIs among various platforms need to be standardized to allow data, transactions, and other information to be exchanged and circulated in various sub-universes, which requires a huge amount of development work. The formation of the metaverse also requires standardized protocols among a series of platforms such as Tencent, Facebook, and Roblox, and the metaverse must comply with the requirements of various national and regional governments.

The metaverse also needs to develop a mechanism to capitalize digital information based on the NFT model, and to form an economic system that allows circulation and trade. NFTs, digital currency (a decentralized example, like Bitcoin, and a centralized example like digital RMB), and real currency must form a complete system of payment, exchange, and withdrawal. Only when a complete standard protocol and economic system are formed can the metaverse really exist.

Although giants such as Tencent and Facebook, as well as gaming companies such as miHoYo, Roblox, and Epic Games, can build several sub-universes on the basis of continuous technical improvement, the sub-universes are separated from each other. This does not form a metaverse, but only a series of highly immersive gaming, social, or industrial internet platforms. The presence of standard protocols and an economic system aggregates these sub-universes into a true metaverse, while these sub-universes remain independent. Only through standard protocols is this possible.

The metaverse will continue to expand as technology advances. Meanwhile, each road will connect to the metaverse system, thus tearing down the boundaries between virtuality and reality.

7.3 The Sub-healthy Metaverse

With the accelerated promotion of hardware and infrastructure, the metaverse prototype has taken shape. However, at present, its ecosystem is in a sub-health state, and there is still a long way to go before a true parallel virtual world is realized. In the future, metaverse development will not only rely on technological innovation to lead, but also on institutional innovation.

Public Opinion Bubbles to be Removed

Under the influence of capital flows, public opinion bubbles can echo stock market volatility. In March 2021, Roblox, an American gaming company, went public, becoming the "first stock of the metaverse." Its price rose 54% on the first day, and its market value exceeded 45 billion dollars, ten times its valuation of 4 billion dollars a year before. This helped the metaverse concept go viral, attracting investors to pour in money.

One month later, Epic Games announced the completion of one billion dollars of financing to create a metaverse space. In July, Mark Zuckerberg announced at the Q2 Facebook financial statement conference that he would assemble a metaverse project team, with the ultimate goal of completely transforming Facebook into a metaverse company in five years. (Facebook has since been rebranded as Meta.) In addition, Nvidia, which has a deep technical accumulation of graphics, takes a fancy to this field. In early August, the company announced that it would team up with Adobe and Blender to make a major expansion of *Omniverse*, planning to open to millions of metaverse users in the future.

Among the Chinese giants, Tencent, ByteDance, NetEase, and Baidu are also loyal followers of the metaverse. As early as 2012, Tencent had targeted this direction and made investments before Roblox went public. Prior to that, it had purchased over 40% of Epic Games shares to create a metaverse ecosystem in all business fields such as social networking, live streaming, and e-commerce.

With an active market, capital has shown great interest in the metaverse. As soon as the news of ByteDance's acquisition of Pico went public, it attracted great attention in the secondary market. Many A-share VR and AR concept stocks have exploded in value. Stocks of Boton Tech and Jinlong Machinery & Electronics hit the 10 percent trading limit in China, and Goertek rose sharply. First-tier funds such as Matrix Partners China, ZhenFund, and 5Y Capital have actively entered the market. For example, 5Y Capital invested in Bolygon game platforms, Party Animal team for individual games, Hyperparameters and X Verse for virtual AI, and invested in Oasis VR for socializing. It covers almost all key areas of the metaverse.

A stream of hot money is pouring into metaverse concept stocks as well. According to statistics from VR Pinea, in June 2021, alone, there were 27 financing mergers and acquisitions in the VR/AR/AI field in China. After the VR studio Recreate Games, (independent from Smartisan) developed the independent game *Party Animal*, which went viral in October last year, investors valued it as hundreds of millions of RMB based on the metaverse concept – this instantly doubled its value. IDC predicted that the global VR products would grow by about 46.2% year-on-year in 2021, and would continue to maintain rapid growth in the next few years; the compound annual growth rate from 2020 to 2024 might reach 48%.

However, despite its present accelerated development trend, the metaverse is still too young when analyzing its actual industrial development. The metaverse industry is still in the foundational stage and far from the ideal state of full industry coverage and ecosystem openness, economic self-consistency, and interconnection between reality and virtuality. The conceptual layout remains focused on XR and game social networking, while the technological and content ecosystems are not yet mature, and the scene entrance needs to be broadened. There is still a long process of "de-bubbling" between the ideal vision and the actual development. It will take at least some years for the market to truly take shape.

For instance, though China already has leading metaverse hardware manufacturers like Pico, the VR content ecosystem, including games, audio, and video, is incomplete. At present, there is only a handful of high-quality VR products good enough to be widely accepted, such as Beat Saber and Half-Life: Alyx. There is a lack of real supreme VR content, which will discourage a large number of users. More importantly, this also involves sensitive issues such as the collection of user privacy data and the establishment of a virtual space social system. These are obstacles in the process of giants building the metaverse ecosystem.

Obviously, creating new concepts, hyping new outlets, and attracting new investments to seek high returns have become commonplace in capital markets: from raising the stock price to reducing holding shares, from conceptual speculation to capital manipulation, from market flattery to regulatory intervention, there are still many uncertainties in the embryonic metaverse; the industry and the market urgently need to regain rationality.

Economic Risks to Be Avoided

In fact, the metaverse economy is the best example of the digital economy. The metaverse is a complete and self-consistent economic system, and the entire chain of pure digital production and consumption. The metaverse economy is no simple industrial revolution, but innovates value creation and redefines the process of value distribution. There is a significant contradiction with the old ideas, the old order, and the old stratum rooted in the traditional economy. To keep up with the pace of change and development of the metaverse, it is necessary to understand its fundamentals more scientifically, avoid possible economic risks, and promote healthy development.

Primarily, the metaverse will operate continuously, and how companies can monetize through it is what must be considered for the long run. In *Roblox* games, players use the virtual currency Robux to purchase access rights to specific games, purchase virtual characters, etc. Roblox, as a platform, takes a cut of the transaction. In this mode, an economic system is formed within these games. Data shows that about 1.27 million developers profited on Roblox in 2020, of which 1,287 made at least 10,000 dollars in the form of virtual currency.

However according to its financial statements, since its establishment in 2004, Roblox remains in a state of fast growth and high losses. Its net loss was $253 million in 2020. Therefore, commercially, how the funds of players, game developers, platforms, and other related parties are divided, and what business model to adopt remain working out. Moreover, in addition to the monetization of socializing, live streaming, games, art, and so on, the metaverse can also complete commercial monetization through intelligent hardware, AI services, digital currency, and ecosystem application stores. No matter what the monetization channel and approach are, business models require further clarification.

Second, there is still a deep tension between the metaverse and the nation-states. Different countries may have different metaverses, while they can create transnational metaverses at the same time; they have a competition/cooperation relationship with each other. Domestically, the government's metaverse strategy is more important, which is a reminder that in the new metaverse era, more comprehensive and proactive measures must be taken to develop the metaverse economy. This means when formulating a metaverse economic strategy, governments should comprehensively consider the potential benefits and obstacles that different policies will bring to economic activity in various fields.

In fact, as for the selection of the development path, there will be many practical difficulties in the process of transitioning from the physical world to the virtual metaverse: the emergence of new things has impacted traditional operation and supervision patterns, and there are numerous institutional barriers. At the technical level, China's foundation is relatively weak, and its technical reserves are insufficient. The current data governance lacks proper means. How to collect, store, manage, and share data elements still needs to be figured out; this is another reminder that for the development of the metaverse economy, a digital strategy that best suits China's national conditions must be formulated. At the government level, the first thing is to strengthen top-level design, as the exploration of development paths requires it to provide corresponding guidance to enterprises; the second is to strengthen digital management and digital legislation; the third is to build a fair and open market environment.

The metaverse economy uses non-exclusive data as production factors, which can break the bottleneck of diminishing marginal effect and provide momentum for sustained economic growth. The development of the metaverse economy is going to enter a new stage. Against such a background, it is necessary to establish a clear and correct understanding of the high multiplier effect of the metaverse economy, so as to accomplish Pareto optimality most suitable for both the metaverse economy and the real economy. Also, to occupy a higher ground in the competition for dimensional upgrading.

At last, though currency and the economic system in the metaverse are not completely linked to the real economy, to a certain extent, they are connected through virtual currency. When the virtual currency in the metaverse world experiences huge fluctuations in value relative to the real currency (legal tender), economic risks will be transferred from the virtual world to the real world. To a certain extent, the metaverse also provides a more subtle manipulation space for the financial harvesting behavior of vast capital; thus, financial supervision also has to expand from the real to the virtual world.

No Laws to Govern the Metaverse

Although the metaverse began with games, it is clearly no game. When new technologies emerge, existing regulations are bound to contradict them. In the cyberspace of the digital age, there are security supervision issues such as user privacy, fraud, viruses, and illegal access to information, which need to be addressed in the metaverse.

(1) To HALT PRIVACY HARVESTING

As a virtual space beyond reality, the metaverse requires fine-grained mining and real-time synchronization of user identity attributes, behavior styles, social relationships, interpersonal interactions, property resources, scenes, and even emotional states and brainwave patterns. This creates higher requirements for the scale, variety, granularity, and timeliness of individual data. The data characteristics of the metaverse era will also react upon personal information and private data, casting a profound impact.

First, there will be an exponential increase in private data in the metaverse era. The metaverse is built on big data, so its data is large-scale and real-time. It will experience significant improvement in data quantity, type, unstructured degree, data collection frequency, real-timeness, and granularity.

Massive data collection supported by multiple technologies will involve more personal and private information, so through the mining and sorting of mass data, user portraits can be easily drawn. Cambridge Analytica's "potential customer" comes from a psychological test app posted on Facebook, which draws a psychological portrait of a person by analyzing social behaviors such as *liking*. There are 5,000 information points on every American, and based on these information points, together with psychological analysis, it suffices to construct an individual's personality model.

By analyzing 10 likes that a person clicks, the algorithm can analyze his/her personality more accurately than his/her colleagues; it only takes 68 likes to estimate the user's skin color (95% accurate), sexual orientation (88% accurate), party affiliation (Republican or Democratic, 85% accurate); with 150 likes, the algorithm can "know" more about people than their parents; with over 300 likes, it will "know" more about people than their spouses.

In addition, data is highly correlated, so privacy is like a domino effect. Applications in many scenarios in the metaverse era are highly dependent on data-related operations. While more value is created, the difficulty of managing private data is greatly increased. Compared with the past, data division in the traditional framework is more obvious, but it also limits private data to a limited scope and department.

In the metaverse era, more data will be connected, and all kinds of data associated with private data are likely to become highly sensitive. Though technologies such as desensitization and de-identification can be employed to deal with them, due to the great number of links, there are more potential threat points of privacy leakage. Scientific and technological progress enables data accuracy and instantaneity to improve in leaps and bounds. Various applications built on this will infiltrate more key areas (such as medical care, health, and finance) related to the national economy and livelihood, while meeting the needs of production, life, and management. Once private data is leaked, there will be serious consequences.

Lastly, data of the metaverse era is also characterized by strong professional processing. With the rapid development of AI, new technologies like deep neural networks have been more widely used. Most AI-based data processing relies on the black-box model, which makes it difficult for non-professionals to understand the data processing process, and easily leads to ethical issues such as data and algorithm discrimination.

The algorithm reveals the imbalance between user data rights and institutional data rights. Data belongs to users, while algorithms to institutions; the collection and use of data are passive for consumers but active for institutions; the algorithms designed by institutions are modelled on their will, so the algorithms endow institutions with great power, and the initiative is always in the hands of institutions.

For institutions, data is transparent. Where there is data, there is an institution. Data belongs to users, but seldom do users know how their data is placed and used – there is an asymmetry between individual data rights and institutional data rights. The characteristic of highly professional data processing inevitably leads to the invasion of individual privacy.

Data privacy is clearly an unavoidable problem from the big data to the metaverse era. Individual privacy data, as the underlying resource that supports the continuous operation of the metaverse, must be constantly updated and expanded. How are these data resources collected, stored, and managed? How to properly authorize and regulate applications? How to avoid data being stolen or misused? How to realize the authentic right and accountability? How to prevent new forms of crime based on data in the metaverse form?

(2) Acknowledge intellectual property

The infringement of intellectual property (IP) has been a "stubborn disease" in the digital space. In fact, in the era of Internet +, the cultural and creative industry has welcomed new development opportunities. However, because of the Internet's network effects, fast transmission, and low cost, there is the endless emergence of various pirate technologies, which cause the industry enormous economic losses.

First, online piracy directly leads to job losses, shrinking royalties, and loss of many excellent works; secondly, the vulgar content of online piracy – together with false advertising, viruses, and low quality of works – gives users a terrible experience, misleads consumer understanding about the real works, and triggers a vicious circle in the copyright market.

Originally, with the rise of the knowledge economy, IP should have become the focal point of the cultural and creative industry. However, IP infringement in the Internet ecosystem has become increasingly serious. There are often lawsuits about network copyright. The problems such as widespread piracy, difficulty in proving IP infringement, and the high cost of safeguarding IP, have become sharp pain spots for the industry.

Norms and technologies are two ways to address legal problems: when the cost of post-lawsuit regulation is high, blockchain is a lower-cost and more efficient alternative. With blockchain, the authentic right of works can be acknowledged through time stamps and hash algorithms, which prove the existence, authenticity and uniqueness of texts, videos, audios, etc. Once that is complete on the blockchain, all subsequent transactions will be recorded in real time, and the entire life cycle of the cultural and creative industry can be traced. This provides a powerful technical guarantee and highly credible evidence for certificates of IP and judicial forensics.

Though blockchain provides technical possibilities for verification, authentic rights, and accountability, massive UGC and IP applications across the virtual and real worlds in the metaverse have exacerbated the complexity and confusion of IP management. The metaverse is a collective shared space, where almost everyone is a creator – this has derived a great number of multi-person collaborative works. There is certain randomness and instability in this collaborative relationship, and explicit rules are needed for such collaborative works and group copyrights.

Elements such as virtual digital people, objects, and scenes in the metaverse are likely to stem or be adapted from corresponding entities in the real world. Such adaptation and applications that cross the virtual-reality boundary are likely to trigger

IP disputes, including portraits, music, pictures, copyright, etc. Virtual humans and objects created by AI in the metaverse may lead to IP disputes too. For example, when a singer performs in the metaverse – is this a form of a commercial concert or online streaming?

When one or more players in a virtual game cooperate to create a virtual item or virtual world, who owns it? Is this product copyrighted? Is it possible to create, protect, or use a brand image in the virtual world? What strategies can content creators deploy to protect their brands in the virtual world? These are important questions for companies that focus on C-end business.

We are on the cusp of the metaverse era, and it is indeed a future worth looking forward to, but its development is not only closely related to the development and maturity of AR, VR, 5G, cloud computing, and other technologies, but its content order, operating mechanism, and so on should undergo multiple public and social discussions.

7.4 New Value Orientation

The ideal metaverse is a utopian-like world that is highly free, highly open, and highly tolerant. As a surreal aggregation of various social relations, complex rules such as moral codes, power structures, distribution logic, and organizational forms need to be explicitly defined and regulated. People have completely different value orientations and beliefs; how to determine the framework of a civilization that supports the metaverse is a complex problem.

A high degree of freedom does not mean unconstrained behavior, and a high degree of openness is not an infinite generalization of boundaries. The social nature of the metaverse is likely to be more complex than reality, and its social system is expected to undergo major changes. As the metaverse develops, the ethics and morality associated with it must develop accordingly; otherwise, disaster will occur. How to build the ethical consensus of the metaverse in a decentralized framework still requires much exploration and thought.

Impacts and Challenges

The advent of the metaverse has definitely impacted human communication and living conditions greatly. The immersion, interactivity, and conception of the metaverse will immerse people in the virtual environment, and they can interact with it, as well as virtual objects, in a natural and real-time manner.

However, it is equally undeniable that the metaverse manifests in an abstract and alternative way. This not only increases the symbolic and virtual nature of people's action environment, but greatly changes the appearance of the human world; thus, challenging the recognition and confirmation of nature, and the overall picture of our world.

In fact, since the 19th century, with the rapid advancement of science and technology and the fast expansion of industrial civilization, the world of human beings has begun to change from natural formation to artificial creation. On the one hand, in our human environment, social organizations based on the primitive connections between people (such as blood and geography), including family, clan, neighborhood, community, village, etc., have been replaced by various artificially constructed social organizations based on purposive corporate actors. On the other hand, in the contemporary human world, the natural environment, such as mountains, rivers, forests, and fields, has also been replaced by artificially constructed material environments, such as skyscrapers and highways.

Despite these two changes, before the advent of the Internet-based metaverse, a certain distance was kept between man and nature; virtuality and reality were clearly distinguishable. However, the development of the Internet and the emergence of the metaverse contain a fundamentally different nature from the various man-made technological innovations of the past.

In the metaverse, human beings have greatly created a greater number of more ethereal and more bizarre symbols. Such symbols not only constrain the world human beings live in more, and blur the relatively clear distinction between the virtual and real worlds, but also challenge the fundamental concepts of "nature" and "reality" in people's life and cognition.

The emergence of the metaverse not only represents the dawn of an era that requires new ways of thinking and acting, but suggests the emergence of a new and different social structure. In other words, before the metaverse integrates reality and virtuality, profoundly altering the place and conditions of human social life, we are destined to face the challenge head on.

New Beginning of the Digital Age

Despite the many theoretical problems and technical obstacles to be tackled in the development of a true metaverse, its huge impact on human life has been visible. The underlying technology clusters that build it increasingly infiltrate various fields, such as science, economy, politics, and culture, thus casting a huge impact on people's social life.

The development of the metaverse is believed to be the inevitable outcome of science and technology advancement. Just as we cannot resist the forces of nature, we cannot deny the existence of the digital age, nor stop its advancement. The future metaverse will not only profoundly affect the ability and methods of humans to understand and transform the world, but quietly change the environment in which people study, work, and live. Therefore, more and more people are becoming accustomed to living and growing in the metaverse; humans will begin to shuttle between the real world and the virtual world.

Its emergence has also had an unprecedented impact on the improvement of cognitive structure and ability of human subjects. The metaverse enriches the cognitive relationship between subject and object, deepens peoples' understanding of this relationship, expands the source of human cognition and knowledge, and provides a new way of living. Therefore, such an emergence will have a significant impact on the traditional cognitive paradigm, and to a large extent, it has caused it to shift already.

The metaverse is a collection of cutting-edge technologies, while the human world is essentially a world of technology. In this world, technology is invented by people; it is a process of human creation and expression. The metaverse is not only revolutionary in how people deal with the relationship between man and the physical world, but also means a lot in the relationships between man and man, and between man and society. It will greatly facilitate people's exploration of complex matters, too.

Therefore, when establishing a new value system, only by linking technology with real lives, and examining the various social factors established by the metaverse as much as possible, can useful guidance on the development direction of the metaverse be provided. In future, we should try to avoid unnecessary waste, and improve efficiency and technological innovation. Ultimately, the great potential of the metaverse should be realized.

Oriented at "Commensalism"

It might be more reasonable and wiser to uphold the idea of "commensalism" (Commensalism is a biological term for a relationship between two organisms, in which one benefits, and the other derives neither benefit nor harm). Furthermore, for the metaverse to become the ladder to happiness for humanity, and for information networking to become a reality that is truly worthy of human longing, we need to establish a concept oriented at commensalism. When we build the metaverse, the commensalism of virtuality and reality is taken as the benchmark when constructing the future world of humans.

(1) BACK TO BENEVOLENCE, RIGHTEOUSNESS, COURTESY, AND FAITH

To establish metaverse ethics, it is plausible to follow the logical starting point of traditional Chinese ethics – "the unity of righteousness and interests."

The "righteousness" is explained in *The Doctrine of the Mean* as suitable. Guan Zi interprets it as being proper everywhere. Such meaning was clarified as early as the Spring and Autumn Period and Warring States periods. *The Commentary of Zuo* contains the words of Duke Zheng Zhuang: "He who is unjust is doomed to destruction." And in the duke's 22nd Year, he said: "it is suitable not to overdrink wine as it is for rituals." *Discourses of the States* (part 2) argues that righteousness follows propriety; therefore, righteousness refers to suitable thoughts and behaviors: "As a moral requirement suitable for propriety, it generally means to have one's own thoughts and actions conform to certain moral standards." It also says that "righteousness is the foundation of interests while greed is the root of resentment. Without righteousness, there is no interest, and with deep greed, there is resentment."

The ethical principle that unites righteousness and interests forms the essence of traditional Chinese ethics, underlining it from start to finish. The relationship between "righteousness" and "interests," in the metaverse era, must reflect the causal connection between moral thoughts and actions, and utilitarian economic behaviors. Establishing the ethical principle of "unity of righteousness and interests" will help create a fine social order in the metaverse, and make it last.

(2) PEOPLE-ORIENTED

In the metaverse, users can enter the virtual space through a user-friendly interface, and when they exit, everything disappears. Consequently, the subject travels seamlessly between the two worlds. People completely determine how and how long the metaverse exists.

In this context, humans will be able to pay more attention to what the metaverse can do for us, and what we can do in the metaverse: how we can make up for our own shortcomings by making full use of its potential; and how to apply human wisdom and intelligence to create a richer metaverse, and form positive feedback. When building the metaverse, we need to stay people-oriented. Only by doing this first can we create a space most suitable for human development.

(3) PRINCIPLE OF SELF-DISCIPLINE

The principle of self-discipline is moral self-discipline characterized by spontaneity. It does not require the supervision and control of the external environment. Self-discipline suffices to keep one following moral norms and moral standards. Jean Piaget, a 20th century Swiss philosopher, believed moral self-discipline arises when the mind can think freely from external pressures. Self-discipline does not mean that there is no rule, no order, or no need for heteronomy. It is gradually formed under the guidance of heteronomy, and it is the result of people's continuous sublimation as they repeatedly practice external codes of conduct.

When building the metaverse, each participant should be self-disciplined, have moral clarity and cognition, and have a clear understanding of the social impact that an individual may cause. For instance, in the gaming industry, today, with the rapid advancement of game development technology, the degree of simulation fidelity grows day by day. Game developers should pay special attention to the impact of violence, blood, pornography, and other content on players, and never use these factors as a gimmick stunt. Meanwhile, they should actively promote "positive energy" in games while pursuing economic interests, such as protecting justice and fighting evil.

7.5 From Reality to Reality

Virtuality is no mystery. It is ultimately a feature of human abstract thinking. In fact, virtuality itself also runs through the development of human civilization and culture.

In the development of human civilization, the first virtuality is physical virtuality. Before the appearance of linguistic symbols, human information was transmitted via physical objects and expressive features of the human body itself. Though physical objects and human bodies are the medium of information transmission, the information they represent and transmit has a "virtual" nature. Physical virtuality uses a certain specific object as a medium to express a certain meaning and information.

Next came symbolic virtuality. Since the appearance of letter symbols, most human information has been transmitted and stored through them, and they have become the material medium. Although the letter symbols used by human beings are a tangible medium, the information transmitted and stored by them is intangible and has a "virtual" nature.

The third virtuality is the digital virtuality that we are experiencing. Digital virtuality relies on Internet technology and certain symbols and images, and its virtual information is digital information converted into computer language. Such information takes the binary form of 0 or 1 as the conversion rule and operation factor, and the bit as the basic unit of information. The digital virtualization of reality is beyond time and space, large, shared and omnidirectional, which makes virtual reality truly possible. And the metaverse completely turns this possibility into reality.

The metaverse exists as a technical cluster in a functional sense. It is an advanced stage of information technology development. But socially, it needs to be pointed out that its generation and development depend on the real world and its development.

On the one hand, the metaverse is not exactly an artificial digital space or a digital mirroring of the real world. As a realistic, non-real thing created by human beings, its non-realness lies in the fact that everything in the metaverse, including itself, is a collection of information rather than a collection of matter. On the other hand, the metaverse can only partially and conditionally mirror matter in the human mind. It is the ultimate form of virtual evolution, and distinguishing its relationship with the real world is of great significance to the development of human society.

Inseparable from the Real World

Whether from the composition of the metaverse or its development, the metaverse relies on the real world, and is only its reproduction and refraction. The real world is always the ultimate reason for the existence of the metaverse.

First, the metaverse is behind the real world in time and logic. In the sense of its technical basis, its realness is supported by real material carriers, such as software, holograms, computer equipment, natural language, sensing means, and pattern recognition.

Although its present technology is still rudimentary, but even if it develops to the final ideal state, the metaverse will still be based on the real world, rather than independent from its material, which is determined by the laws of human survival and development. In a broader sense, the extent to which humans understand and transform the real world determines how far the metaverse goes.

Digital virtuality directly reflects the human ability to understand and transform the real world, as well as the limitations on making those happen. It is an endless journey for us to understand and transform the real objective world; correspondingly, so is the development of digital virtualization.

Secondly, as a simulated world constructed by people with the help of modern science and technology, the metaverse resembles direct human existences in the real world. Fundamentally, humans are the subject of the real world first and the metaverse next, having first built the real world and then the virtual one.

It is on this fundamental point that the real world and the metaverse are connected:

the subjects in both worlds are real people – people transform the real world according to their own needs and interests, which is also the case in constructing the metaverse. Therefore, both worlds are the realization and manifestation of human imagination and power.

For the metaverse, as people construct and operate it, the sense of reality caused by the things and situations in it must be based on the reality of the subject; the technical equipment of the metaverse must be worn by a real subject (person) to be effective. And the multi-sensory signals of vision, hearing, smell, and touch generated during people's interaction with objects in the metaverse can only be recognized by real people.

The strong dependence of the metaverse on people in the real world determines that its generation must rely on reality; its objects originate from the real world. As a way for human beings to express the real world, the particularity of the metaverse is manifested in the use of the binary digital expression to perform simulations.

Simulation is necessarily inseparable from real-world objects: the metaverse can generate a sensory experience similar to the real world – one that can be confused with real world objects in terms of vision, hearing, smell, taste, and touch – because it is connected and consistent with the way people perceive the real world.

However, virtual objects in the metaverse are created on the basis of information material from real objects, and are properly processed by humans on this. Without the existence of real objects, it is impossible to create virtual ones. Clearly, if it is not closely related to the real world, the metaverse can only provide the human/subject with an unfamiliar realm, and those who go deep inside will be as dazed as computer game beginners, sinking into a state of confusion.

Metaverse Prosperity Depends on Real World Development

The metaverse is the outcome of information technology development to a certain stage. Without the development and maturity of 5G, cloud computing, AI, VR/AR, and other technologies, the metaverse would only exist in science fiction. In other words, for it to develop further, continuous innovation of science and technology in the real world is essential. The advancement of technology will affect not only all aspects of the metaverse in practical applications, but also its extensive application in other new areas.

Real world development is also the ultimate destination for metaverse development. This is because, in the metaverse, people interact with virtual objects as subjects, and the changes in virtual objects caused by this interaction do not correspond to the changes in real-world objects; they are only a simulated demonstration of transformation activities.

For the metaverse to play a real role, people need to fully absorb the experience gained within it and apply this to real world transformation. In other words, real world transformation can never be completed in the metaverse. Therefore, fundamentally, the development of the real world determines the development of virtual reality, and the development of metaverse ultimately serves the development of the real world.

Going Beyond the Real World

In the metaverse, there is digital existence, which can break through the limitations of time and space in the real world. It makes what is impossible in reality possible, thus going beyond the real world.

First, in the traditional physical world, all human activities are bound to be restricted by social norms and practices. Humanity strives to create an ideal state that meets human requirements. And VR technology is the contemporary manifestation of this pursuit.

In previous societies, individuals were usually submerged in the collective. This was especially true in capitalist societies which emerged from industrial civilization, with production practices standardized and modeled. The purpose of production is to obtain as much profit as possible, while individual needs are often neglected. In this case, individuals can usually act only as passive recipients, and rarely have the opportunity and rights to make independent choices. Individuality is often obscured by the wholeness of society, but digital existence provides a space for us humanity to pursue this.

In the metaverse, people are no longer controlled and impacted by the physical world, and individual freedom and independence have become possible. In a way, human individuality is maximized in the metaverse. In such an environment created by digitization, everyone is the master and can choose freely, which fully reflects the autonomy and independence of the subject. This is of great significance for the free and comprehensive development of humanity.

The metaverse provides a new opportunity for freedom. People can be liberated from the daily mundane state of existence that is constrained by various pressures; they can enter a free state of existence to some extent, alleviate stress from laws and other real-world conditions, and stop worrying about age, race, gender, identity, and so on – they can simply exist as independent individuals.

In the virtual space, people can even set themselves various roles, choose the life they are interested in, express their thoughts and emotions freely, show the part of themselves they want to be seen, and reach emancipation. It is in this direct interaction with such a world that the meaning of life – which is usually obscured and suspended in limited utilitarian activities – will truly be questioned. The subject utilizes all intellect and emotion to embrace this meaning, gaining a whole new world.

In this world, life is highly free, the mind is liberated like never before, and the subject experiences a kind of intoxicated pleasure. It is in this sense that the essence of metaverse will transcend technology and become art: the metaverse does not control, escape, entertain or communicate, but ultimately changes and remedies our sense of reality.

Metaverse Development Promotes the Real World

The communist society as envisioned by Karl Marx was a beautiful prediction for the ultimate evolution of human society. Essentially, it was people-oriented and aimed to reach harmony in all aspects. Meanwhile, it is also a most scientific and harmonious society, where everyone enjoys dignity, fairness, and justice, and where human needs for a better life continue to be met and improved. Also, various social ills, including alienation and mass inequality, are eliminated, responsibilities and rights form the benchmark, and cooperation and sharing are widespread.

Marx predicted that in a communist society, nobody would be a specialist as such, but instead, they could develop in any sector. Society regulates the entire production process, thus making it possible for us to do whatever interests us at all times. We can hunt in the morning, fish in the afternoon, garden in the evening, and take part in debates after dinner – so we won't always be a hunter, fisherman, gardener, or critic. In a communist society, there is the free development of all people.

Yet in reality, such ideal freedom requires the corresponding material and technical foundations to happen in real life. Due to the limitation of realistic conditions, it is impossible for us to have all the kinds of things we wish in contemporary society. Our study, work and life are often limited to a certain range, follow a fixed pattern, and have a single form. They are not as rich as Marx predicted.

However, the emergence of virtual reality provides us with such a possibility. People do not have to have fixed occupations. As long as they are immersed in set metaverse scenes, they can experience the tension and excitement of a great many activities, which have been realized in many of the latest computer games.

Though the metaverse is no real scene, the subjective experience it brings to people is real, effective, and rather impressive. At present, with the continuous development of VR, virtuality and network communication have gradually improved and will be further integrated. In cyberspace, people can communicate and exchange regardless of their identity, status, blood ties, region, and ethnicity. Internet communication has significantly broadened people's communication space and increased the frequency of interpersonal communication.

Traditional network communication, limited to computer screens, is far from face-to face communication in real life. When VR is combined with network communication, new methods have come into being. Such real-time communication has the potential to provide endless opportunities for people of different cultures to exchange ideas and pool their collective intelligence.

The promotion of the metaverse is reflected in the application of VR in human society, too. In the future, the metaverse will also be applied to social fields such as the military, medical care, leisure, and entertainment. In medicine, it can be used for various simulations, including human anatomy simulations for medical students, enabling them to practice surgeries with the help of virtual patients; in the military field, it can provide commanders with military drill scenarios, and pilots with simulated flight training; in urban planning, VR can be applied through urban simulation: people can place various planning schemes in the virtual environment, consider their impact on the real ecological environment, and evaluate various options. Users can also judge the rationality of space design, thus avoiding huge capital and time wastage in actual construction.

As the times roll forward, the metaverse will also exhibit broader and better features, promote social development, effectively propel humanity, and help us truly realize the wonderful communist society as Marx predicted.

Postscript

The metaverse is without doubt inevitable. It is a new form of business that will definitely occur considering humanity's enormous technological developments. At present, whether the metaverse is immature or ready for a capital investment boom, we shall remain calm. Like the IoT boom 10 years ago and the smart wearables boom 5 years ago, the industry's promising future does not mean everything will fall into place in one go – it takes time to grow. The metaverse is not the fruit of a single technological breakthrough, but the superposition of multiple technologies.

If I am to give my own prediction, the total arrival of the metaverse will be after 2040. Therefore, leading enterprises must attach great importance to studying, following, investing in, and developing related industrial technologies. Common entrepreneurs need to concentrate their advantageous resources as much as possible on a small piece of the industry chain, and persistently develop and iterate. As the industry chain grows and matures, these entrepreneurs are going to hold a seat at the industry chain on the day the metaverse era comes.

At last, after finishing this book, if you do gain a clear understanding of the metaverse, please remember my personal definition of it:

The so-called metaverse is a product driven by a variety of technologies; a mixed world of virtuality and reality. Both individuals and the physical world will become ubiquitous and accessible because of such technology. Meta is used to suggest a new epoch – something with unknown boundaries, and verse is used to depict the vastness of the coming mixed world. We call it metaverse because when the virtual world and the real world are intertwined, their boundary becomes unknown.

Since it is formed from the amalgamation of technologies, the lagging development of any link in the entire industrial chain will impede the boom of the metaverse. However, it is an era that is bound to arrive, because the technology is constantly advancing.

When the metaverse era truly comes, our current way of life, business modes, social supervision, and governance patterns, including ethics and value systems, will be reshaped. Our body will not be immortal in that era, but our thinking will – with the fusion of artificial intelligence. In such an era, everything will be redefined, and what we need to do is embrace greater change with a more open mind.

References

CITIC Construction Investment Securities. "Special Report on Metaverse: Begins with Games, But More."

CITIC Securities. "RBLX Investment Value Analysis Report: Roblox – the World's Leading Multiplayer Game Creation and Social Platform."

CITIC Securities. "Special Research Report on Metaverse: Begins with Experience, and Breaks the Boundary of Virtuality and Reality."

CITIC Securities. "The 177-page In-depth Report on Metaverse: The Digital Survival of Human Beings, Entering the Early Exploration Period."

Guosen Securities. "Economic Research Institute. Special Research Report on Metaverse: A New Epoch of Cyberspace."

Guosheng Securities. "Collaborative In-depth Research on Metaverse and Blockchain Industry: Metaverse, the Next Stop of the Internet."

Guosheng Securities. "The Metaverse of the Blockchain Industry: Reconstruction of Computing Power, the Ladder to Metaverse."

Guosheng Securities. VR Industry Study: VR on the Rise Again, and Its Wide Application.

REFERENCES

Hu, Xiaoan. "Research on Some Philosophical Issues of Virtual Technologies." Wuhan University, 2006.

Huaan Securities. "In-depth Report on Metaverse: Is the Metaverse the Ultimate Form of the Internet?"

Huatai Research. "Prospects of the Show Business in 2030: Embrace the Digital Life That Combines Virtuality and Reality."

Huawei. "White Book of AR Insights and Application Practice."

Huaxi Securities. "In-depth Report on Metaverse: the Next 'Ecological Level' Technical Main Thread."

New Media Research Center of Tsinghua University. "2021 Metaverse Development Research Report."

Orient Securities. "Special Report on Blockchain: NFT, New Elements of User Ecology, The Potential Economic Carrier of Metaverse."

Tianfeng Securities. "In-depth Research Report on Metaverse: Building a Virtual Digital Second World Independent of the Real World."

Tianfeng Securities. "Special Research Report on Roblox: the First Share of Metaverse, Leader of Metaverse."

Zhao, Yangyang. "An Ethical Review of Online Game Development." University of South China, 2013.

ZTE. "5G Cloud XR Application White Paper 2019."

Index

ABOUT THE AUTHOR

Kevin Chen is a renowned science and technology writer and scholar. He was a visiting scholar at Columbia University, a postdoctoral scholar at Cambridge, and an invited course professor at Peking University. He has served as a special commentator and columnist for the *People's Daily*, CCTV, the China Business Network, SINA, NetEase, and many other media outlets. He has published monographs in numerous domains, including finance, science and technology, real estate, medical treatments, and industrial design. He currently lives in Hong Kong.